Nelson Science

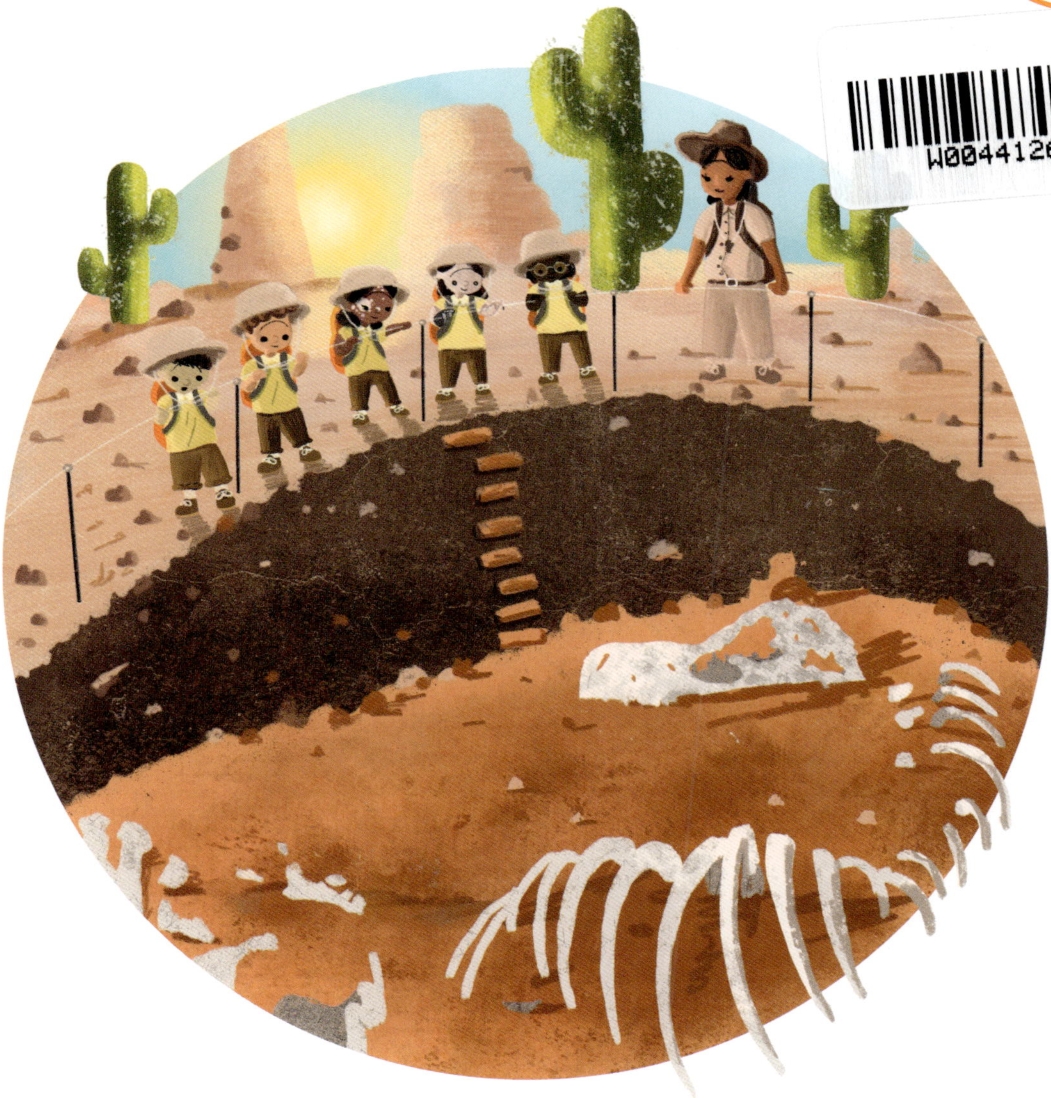

Pupil Book

Anthony Russell

OXFORD
UNIVERSITY PRESS

OXFORD
UNIVERSITY PRESS

Great Clarendon Street, Oxford, OX2 6DP, United Kingdom

Oxford University Press is a department of the University of Oxford.

It furthers the University's objective of excellence in research, scholarship, and education by publishing worldwide. Oxford is a registered trade mark of Oxford University Press in the UK and in certain other countries.

British Library Cataloguing in Publication Data

Data available

ISBN: 978-1-382-01748-0
ISBN: 978-1-382-01749-7 (Pupil book only)

1 3 5 7 9 10 8 6 4 2

Paper used in the production of this book is a natural, recyclable product made from wood grown in sustainable forests. The manufacturing process conforms to the environmental regulations of the country of origin.

Printed in Great Britain by Bell and Bain Ltd, Glasgow

Acknowledgements

The publisher and authors would like to thank the following for permission to use photographs and other copyright material:

Cover: Aaron Cushley. **Photos: p1(a):** The Picture Studio/Shutterstock; **p1(b):** Ria Yui/Shutterstock; **p1(c):** AlanMorris/Shutterstock; **p1(d):** Shutterstock/Victor Tyakht; **p1(e):** FloridaStock/Shutterstock; **p1(f):** Volodymyr Burdiak/Shutterstock; **p1(g):** WildMedia/Shutterstock; **p1(h):** Rclassen/123RF; **p1(i):** Henrik Larsson/Shutterstock; **p2(tl):** Ami Parikh/Shutterstock; **p2(tr):** Mario Krpan/Shutterstock; **p2(ml):** Michael Potter11/Shutterstock; **p2(mr):** Leungchopan/Shutterstock; **p2(bl):** Igor Kovalchuk/Shutterstock; **p2(br):** Paulinux/Shutterstock; **p3(l):** Garfotos/Alamy Stock Photo; **p3(r):** Keattikorn/Shutterstock; **p4(l):** LuciaP/Shutterstock; **p4(r):** Ihor Hvozdetskyi/Shutterstock; **p5(tl):** imageBROKER/Alamy Stock Photo; **p5(tm):** Peter Martin Rhind/Alamy Stock Photo; **p5(tr):** Paul Reeves Photography/Shutterstock; **p5(b):** Nella/Shutterstock; **p6(tl):** Alekuwka/Shutterstock; **p6(m):** Apiguide/Shutterstock; **p6(tr):** carlosobriganti/123RF; **p6(bl):** AB Photographie/Shutterstock; **p6(br):** Rob Hainer/Shutterstock; **p7(tl):** Rclassen/123RF; **p7(tr):** Divelvanov/Shutterstock; **p7(m):** Mario Krpan/Shutterstock; **p7(bl):** Roaming Panda Photos/Shutterstock; **p7(br):** Henner Damke/Shutterstock; **p8(tl):** Medtech THAI STUDIO LAB 249/Shutterstock; **p8(tr):** Kateryna Kon/Shutterstock; **p8(bl):** Lebendkulturen.de/Shutterstock; **p8(br):** Lebendkulturen.de/Shutterstock; **p9(t):** Ed Reschke/Photodisc/Getty Images; **p9(mr):** InsectWorld/Shutterstock; **p9(b):** Leena Robinson; **p9(ml):** Neo Edmund/Shutterstock; **p10(a):** Pius Rino Pungkiawan/Shutterstock; **p10(b):** Trevor Keyler/Corbis; **p10(c):** Mitch Shark/Shutterstock; **p10(d):** HWall/Shutterstock; **p10(e):** Blik Sergey/Shutterstock; **p10(f):** Ole Schoener/Shutterstock; **p11(l):** Valsib/Shutterstock; **p11(r):** Vesna Kriznar/Shutterstock; **p12(a):** Lubov62/Shutterstock; **p12(b):** Justin Bezuidenhout/Shutterstock; **p12(c):** Arzu Kerimli/Shutterstock; **p12(d):** luis2499/Shutterstock; **p12(e):** Carmine Arienzo/Shutterstock; **p12(f):** Vinicius Tupinamba/Shutterstock; **p12(g):** Frank60/Shutterstock; **p12(h):** Vilainecrevette/Shutterstock; **p12(i):** Chris Mattison/Alamy Stock Photo; **p12(j):** Tomas Drahos/Shutterstock; **p12(k):** All Canada Photos/Alamy Stock Photo; **p12(l):** Matthew Williams-Ellis/Alamy Stock Photo; **p12(m):** Crystaltmc/Shutterstock; **p12(n):** Gsplanet/Shutterstock; **p12(o):** Nir Darom/Shutterstock; **p12(p):** Craig Lambert Photography/Shutterstock; **p14(l):** Cristian Zamfir/Shutterstock; **p14(r):** Grzegorz Firlit/Shutterstock; **p18:** Mix Tape/Shutterstock; **p19:** Fotosr52/Shutterstock; **p21:** Fotokostic/Shutterstock; **p22(t):** Hi Brow Arabia/Alamy Stock Photo; **p22(b):** Sportpoint/Shutterstock; **p25:** Gregory Johnston/Shutterstock; **p26:** Maurice Savage/Alamy Stock Photo; **p27(tl):** Gabriel12/Shutterstock; **p27(tm):** MikeBlack/Shutterstock; **p27(tr):** Cathy Keifer/Shutterstock; **p27(bl):** YuliaTroizk/Shutterstock; **p27(br):** Natalia Golovina/Shutterstock; **p28(tl):** Vinicius Tupinamba/Shutterstock; **p28(tr):** Oleksandr Ieremenko/Shutterstock; **p28(bl):** Stewart Innes/Shutterstock; **p28(br):** Steven Ellingson/Shutterstock; **p30(tl):** Lopolo/Shutterstock; **p30(tr):** PedkoAnton/Shutterstock; **p30(b):** Catherine Delahaye/DigitalVision/Getty Images; **p32:** zhanglianxun/Shutterstock; **p35:** Kumar Sriskandan/Alamy Stock Photo; **p36:** Pavel Chagochkin/Shutterstock; **p37:** TY Lim/Shutterstock; **p38:** Dmitry Kalinovsky/Shutterstock; **p40(a):** alice-photo/Shutterstock; **p40(b):** AlessandroZocc/Shutterstock; **p40(c):** wannasak saetia/Shutterstock; **p40(d):** Natalia van D/Shutterstock; **p40(e):** Steffen Foerster/123RF; **p44(tl):** Jimmy Rubisky/Shutterstock; **p44(tm):** Worldswildlifewonders/Shutterstock; **p44(tr):** Blickwinkel/Alamy Stock Photo; **p44(bl):** Al Qaralleh/Shutterstock; **p44(bm):** Cbstockfoto/Alamy Stock Photo; **p44(br):** Serge Vero/Shutterstock; **p45(l):** Vladimir Melnik/Shutterstock; **p45(r):** Idiz/Shutterstock; **p50(l):** dpa picture alliance/Alamy Stock Photo; **p50(r):** AuntSpray/Shutterstock; **p51(l):** Evgenii mitroshin/Shutterstock; **p51(r):** Monkey Business Images/Shutterstock; **p54:** Gatot Adri/Shutterstock; **p59:** Alexandra Jursova/Moment/Getty Images; **p60:** shaineast/Shutterstock; **p64(a):** SCIENCE PHOTO LIBRARY/Getty Images; **p64(b):** Vandrage Artist/Shutterstock; **p64(c):** Art Directors & TRIP/Alamy Stock Photo; **p64(d):** Anttoniart/Shutterstock; **p64(e):** Studio217/Shutterstock; **p72:** Fototocam/Shutterstock; **p75(l):** David J. Green/Alamy Stock Photo; **p75(m):** David J. Green/Alamy Stock Photo; **p75(r):** Powered by Light/Alan Spencer/Alamy Stock Photo; **p76:** SAPhotog/Shutterstock; **p77(tl):** Zelenskaya/Shutterstock; **p77(tr):** TuktaBaby/Shutterstock; **p77(bl):** Vvoe/Shutterstock; **p77(br):** Vvoe/Shutterstock; **p78(l):** Aleksandr Pobedimskiy/Shutterstock; **p78(r):** Warunee Chanopas/Shutterstock; **p79(tl):** Valery Voennyy/Alamy Stock Photo; **p79(tr):** Fokin Oleg/Shutterstock; **p79(b):** P Tomlins/Alamy Stock Photo; **p80(tl):** Tyler Boyes/Shutterstock; **p80(tr):** www.sandatlas.org/Shutterstock; **p80(bl):** Vvoe/Shutterstock; **p80(br):** TuktaBaby/Shutterstock; **p82(t):** Drug Naroda/Shutterstock; **p82(m):** Q Studio Travel/Alamy Stock Photo; **p82(b):** Richard Semik/Shutterstock; **p83(t):** Hadrian/Shutterstock; **p83(m):** Fotos593/Shutterstock; **p83(b):** Doug Perrine/Alamy Stock Photo; **p89(r):** Ratikova/Shutterstock; **p89(l):** Filippo giuliani/Shutterstock; **p89(m):** Mckie/Shutterstock.

Artwork by Q2A Media Services Pvt. Ltd., Maurizio de Angelis, Tony Forbes, Simon Rumble, and Wearset Ltd.

Every effort has been made to contact copyright holders of material reproduced in this book. Any omissions will be rectified in subsequent printings if notice is given to the publisher.

Contents

UNIT 1 Living things and their habitats

Classification of living things

Plants and animals

In Year 4 you learnt that living things can be sorted and grouped in different ways.

You divided the living things into two groups – **plants** and **animals**. You then divided each group into smaller sub-groups.

Plants

Trees

Shrubs

Herbs

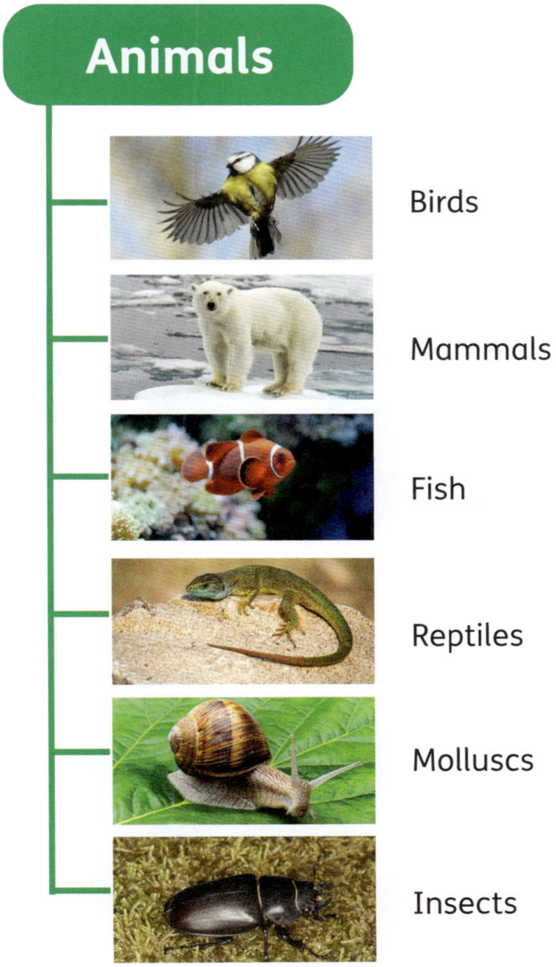

Animals

Birds

Mammals

Fish

Reptiles

Molluscs

Insects

Can you remember what features the plants and animals in each group have?
Talk about it with your class. If you need a reminder, look back at Unit 1 of the Year 4 Pupil Book.

➡ *Workbook page 1*

When you sort things into groups, you **classify** them. The chart on the previous page reminds you of one way to classify living things.

You can classify living things in other ways too.

• You can classify animals by whether or not they have a backbone – the **vertebrates** and the **invertebrates**.

• You can classify animals by what they eat – the **carnivores** (meat eaters), the **herbivores** (plant eaters) and the **omnivores** (plant and meat eaters).

• You can classify the **producers** in a **food chain** – (those that produce food) and the **consumers** (those that eat food).

• You can classify plants by whether they have flowers or not.

Classifying flowering plants

There are many different kinds of flowering plant, so this group is very large. It is useful to divide this group into smaller sub-groups, based on the different features of the flowers.

Flowers have different colours and sizes, but they also have differences in the parts within them. **Petals**, **sepals**, **stamens** and **stigmas** are not the same in all flowers. Their number and their shapes differ. These different characteristics of the flowers help us to classify the flowering plants in smaller groups.

Can you think of any other ways to classify living things?

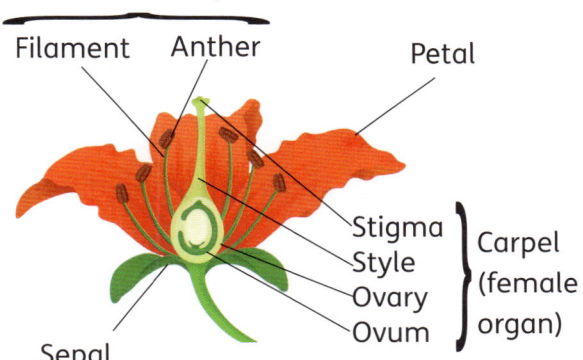

Stamen (male organ)

Filament   Anther

Petal

Stigma
Style
Ovary
Ovum

Carpel (female organ)

Sepal

Activity 1

You will need: flowers from outside or from the room, hand lens, paper (or Workbook) and a pen or pencil.

1 Go outside and look for flowers that you can pick. Choose at least three different flowers, with different shapes if possible.

You must only pick the flowers that your teacher tells you are OK to pick.

2 Look closely at each flower.

Make a drawing of each flower before you begin to take pieces off it.

3 Now **record** details about the petals, sepals, anthers and stigmas of each flower.

> You may need to gently pull parts off the flower in order to see everything.

   a Draw a table like the one below.

   b Count the number of each part and write the total in the table.

   c Draw each part in the table, taking care to show the size and shape. Colour your drawings.

Flower	Petals	Sepals	Stamens	Stigmas

4 Look at your **results** and make comparisons between the parts of the different flowers.

Look for similarities and differences.

5 How could you group the flowers using your **observations**?

   a Discuss your ideas with a partner and come to a conclusion.

   b Share your groupings with the class. Listen to the ways in which other people grouped their flowers.

Look at the pictures of the five flowers and identify similarities. Tell the class what you **observe**. Explain why you think they have been classified in the same family.

Wild rose

Apple

Almond

Strawberry

Cinquefoil

Some of the rose family

Botanists have classified all these plants as members of the **rose** family. Their scientific name is Rosaceae. There are nearly 5000 different **species** of plant in this family and they are grouped into 91 different groups!

All plants in the rose family have flowers with five petals and five sepals and usually many stamens. You can see the petals in the pictures above and they are easy to count.

The roses we grow in our gardens and buy from shops are not natural roses. They have been bred by plant breeders to have many more petals, so they look very different. But they are still members of the rose family.

This is one example of how living things can be grouped into sub-groupings. These sub-groupings are often called families (like the rose family).

All the members of family must have similar characteristics (like all having five petals). This is the family identity.

Classifying cats

Zoologists also group animals into sub-groups called families.

cat

lynx

tiger

Some of the Felidae family

black panther

The animals in the pictures above are all mammals, but they have been classified in a smaller family called 'the cats' or 'Felidae' because they share important characteristics. Look at the pictures and try to work out what some of these characteristics may be. Tell the class what you think.

If you have a pet cat you will have discovered some of the reasons why it is in the 'Felidae' family. All members of this family have 30 teeth, eat meat, are **predators**, have claws that can be drawn in, have five toes on the front feet and four on the back feet and the pads of their feet are in three cushions. There are 31 species in the cat family, divided into 12 smaller sub-groups.

lion

Classifying invertebrates

There are lots of different invertebrates (animals without a backbone). We can divide them into smaller groups, using their characteristics to identify different families of invertebrates.

For example, the **mollusc** family is one of the largest families of invertebrates. Molluscs come in different sizes, shapes and structures. Zoologists have found over 85,000 species of mollusc! Most of them live in water and feed on plants. The one characteristic they all share is a body without **segments**. Most also have a shell.

This makes them different to the worm family, which are also invertebrates, but they have segments in their bodies.

snail

squid

slug

oyster

octopus

Some of the mollusc family

The giant squid is a mollusc and is also the largest invertebrate on Earth. These animals can measure up to 18 metres long! They hunt and catch other animals to eat.

We all know much smaller members of the mollusc family that don't live in water. The slugs and garden snails are plant eaters that live on land. Slugs and snails look different on the outside – snails have shells and slugs don't – but they can still be part of the mollusc family because they don't have segments. It's often *not* the visible characteristics that link animals in the same family.

Families of micro-organisms

Micro-organisms are tiny living things that are too small to be seen without a microscope. These are not like plants and animals, which are linked by how they feed. The only thing that links the wide variety of micro-organisms is that they are **microscopic** – which gives them their name.

We can sort and classify micro-organisms.

Bacteria have different shapes and we classify them by their shape. **Viruses** are in a group by themselves. Many other simple organisms with only one cell (called **protista** or **protozoa** or **parasites**) live in water and the soil and they are very important in helping things decay.

Some protozoa cause diseases in humans and animals (for example the malaria parasite or the sleeping sickness parasite).

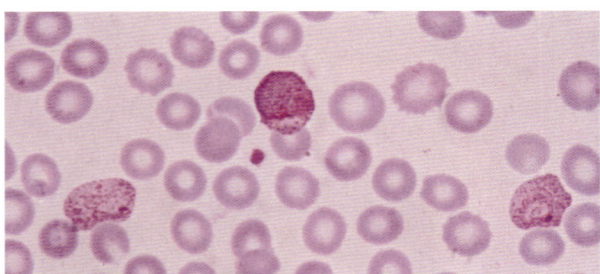
malaria parasite

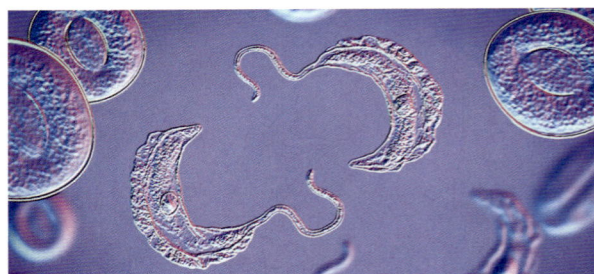
sleeping sickness parasite

euglena

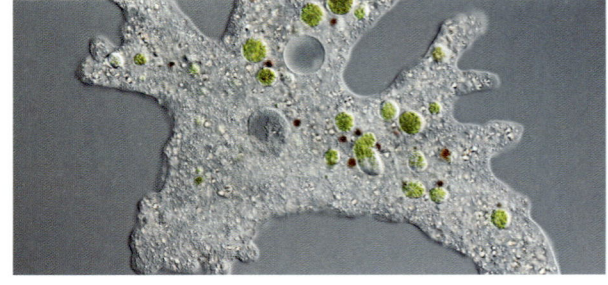
amoeba

Characteristics used in classification

Biologists have sorted living things into four main groups: the protista, the **fungi**, the plants and the animals.

The one big difference between plants and animals is how they feed. Plants make their own food from sunlight through the process of **photosynthesis**. Animals cannot do this. They eat living or dead plants, or animals, or both.

This difference between plants and animals has nothing to do with what either of them looks like from the outside. Animals can be all different shapes and sizes, but a living thing belongs to the animal group if it eats other living things.

Plants can be all sorts of colours and shapes, but a living thing belongs to the plant group if it makes its own food from sunlight.

Fungi cannot make their own food. Fungi feed on living or dead organisms. They are either parasites (for example mildew or athlete's foot) or **saprophytes** (for example mushrooms, or moulds such as Penicillium).

Once an organism has been allocated to one of the four main groups, scientists look for other characteristics to help divide the group into smaller families. Here are two families you have already learnt about.

- Rosaceae are a family in the plant group.
  All members of the Rosaceae family are flowering plants where each flower has five petals.

- Molluscs are a family in the animal group.
  They are all invertebrates and their bodies don't have segments.

The **life cycle** of an animal is very useful in deciding which family to put it in. For example, you learnt in Year 5 that a caterpillar changes into a butterfly during its life cycle.

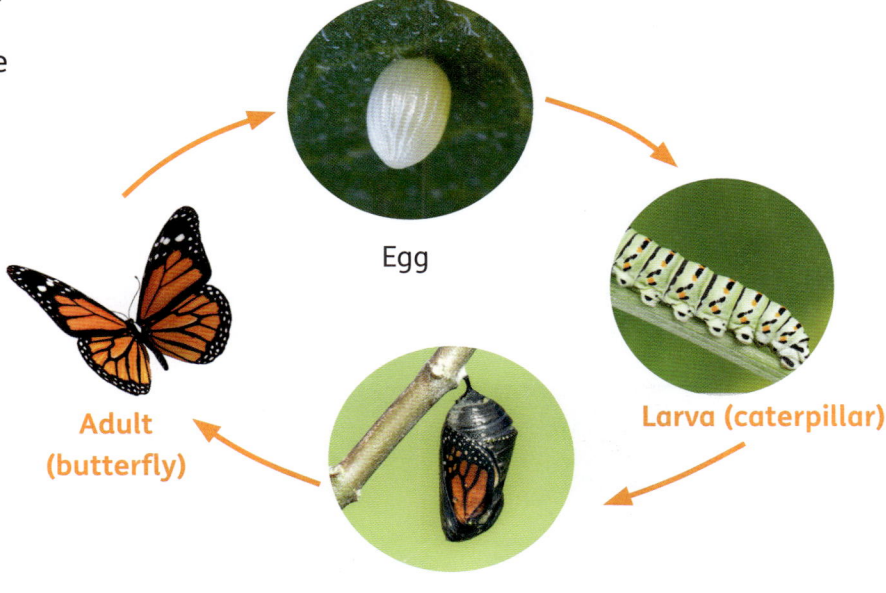

Egg

Larva (caterpillar)

Pupa (chrysalis)

Adult (butterfly)

Insects are the *only* kind of animal that completely change their physical shape in this way – so, you know that a caterpillar must be in the insect family.

Are there other families that the caterpillar belongs to if you look at other ways to classify animals?

Activity 2

You will need: paper (or Workbook) and a pen or pencil.

crocodile

lizard

salamander

newt

Are they in the same family?

1 Look at the four animals above and discuss with your group how you could classify them. How could you classify them so that they are:

 a all in the same family **b** not in the same family?

 Explain your reasoning.

Some of the physical differences between plants are less obvious. Scientists have to look more closely at several features of plants to classify them.

datura

campanula

convolvus

lily

2 Look at the four flowers and discuss with your group how you could classify them. How could you classify them so that they are:

a all in the same family

b not in the same family?

Explain your reasoning.

Look again at the crocodile, lizard, salamander and newt. If you classify them by how they look, you could easily put them all in the same class. They all have four legs, a head, a tail and they all breathe air though nose and mouth.

However, scientists don't put them in the same class. Newts and salamanders are amphibians – they lay their eggs in the water, and don't have scales.

Crocodiles and lizards are reptiles – they lay eggs on the land and they have scales on their bodies.

Look again at the different flowers in the activity. The flowers are all from flowering plants, they are all trumpet-shaped and they are all colourful, but botanists put them in four different families. So why were all these flowers not put in one family?

The flower shape is not the basis of a 'family'. The number and arrangement of the flower parts is much more important. Roses have five petals and five sepals and many stamens. Other plants with trumpet-shaped flowers do not have the same number of petals, sepals and stamens as roses – only the shape of the flower is common to all of them.

These examples show you that scientists rarely classify living things by what they *look like* on the outside.

Activity 3

You will need: paper (or Workbook) and a pen or pencil.

1 Look at the pictures and try to name them. Tell your group what you think.

(a)

(b)

(c)

(d)

(e)

(f)

(g)

(h)

(i)

(j)

(k)

(l)

(m)

(n)

(o)

(p)

2 If there are any animals that your group cannot name, ask your teacher to tell you their names.

3 Sort the animals into two sets.

Try not to classify the animals by simple things like what they look like or where they live, but try to use a scientific **classification**.

  4 Share your sets with the class.

    a Did others group the animals in the same way as you did?

    b Discuss any differences in the way you grouped them.

You probably realised that it was best to group the animals in Activity 3 by whether they were vertebrates or invertebrates. All the animals in the first set have a backbone.

The backbone runs down the back of the body, in the centre. Vertebrates all have a skull attached to the backbone. They have a **brain** inside the skull and there are **sense organs** that send information about the outside world to the brain through the **nervous system**.

The majority of vertebrates have four limbs that they use to move. These can be legs, wings, arms or fins and they are used for walking, flying, swimming, swinging, jumping, climbing or running. Vertebrates also have a heart that pumps blood around their bodies.

All the other animals on Earth are classified as invertebrates. This means they do not have a backbone. Look again at the animals in the second set in Activity 3. You can see that these invertebrates are very varied in form and structure.

• Some have limbs and others do not.

• Some have a skeleton on the outside and others do not.

• Some have bodies divided into parts or sections and others have no different sections.

• Some have a head and others do not.

There is only one common characteristic that all invertebrates have – they do not have a backbone!

Science in Action

The modern system of classification was invented by a Swedish botanist called Carl Linnaeus (1707–1778). He was not happy with the way biologists used several words to identify each species of living thing. It was a complicated and 'clumsy' way of naming living things.

Instead, by using earlier ideas from other scientists, he published his simpler system, which gives two names for each organism. This is called the 'binomial system' and it is still used today by scientists all over the world. Each species discovered by explorers and other researchers is given two Latin names, which help identify how it is classified.

At the same time, living things are given local names in each country's language. These are more commonly used in everyday life.

The Natural History Museum in London announced that there were 503 new species discovered in 2020 and we can expect to add more to the list as our exploration of the oceans and more remote **habitats** on land continue in the future.

Popa langur
(Trachypithecus popa)

the hairy worm
(Oligobregma brasierae)

Two of the 503 new species found in 2020

End of unit check

Choose the correct answer to complete these statements.

1 Vertebrates all have:

    a a backbone, scales, cold blood

    b warm blood, a backbone, two limbs

    c feathers, a backbone, cold blood

    d a skull, a backbone, a brain

2 The following animals are invertebrates:

    a snail, whale, mosquito

    b mosquito, snake, cockroach

    c fish, snail, bee

    d cockroach, mosquito, snail

3 Humans are classified as mammals because:

    a we have two legs and are warm blooded

    b we give birth to live young and have a backbone

    c we have a backbone and a skull

    d we give birth to live young and mothers feed them with their milk

4 These are examples of protozoa:

    a fungi, virus, algae

    b amoeba, fungus, bacteria

    c euglena, amoeba, malaria parasite

    d virus, algae, amoeba

5 The group of plants are all living things that:

    a do not move around

    b have roots

    c make their own food from sunlight

    d have colourful flowers

6 Crocodiles and salamanders are *not* in the same class because:

    a one lays eggs in water and the other on the land

    b they live in different parts of the world

    c crocodiles are much bigger than salamanders

    d they eat different foods

The human circulatory system

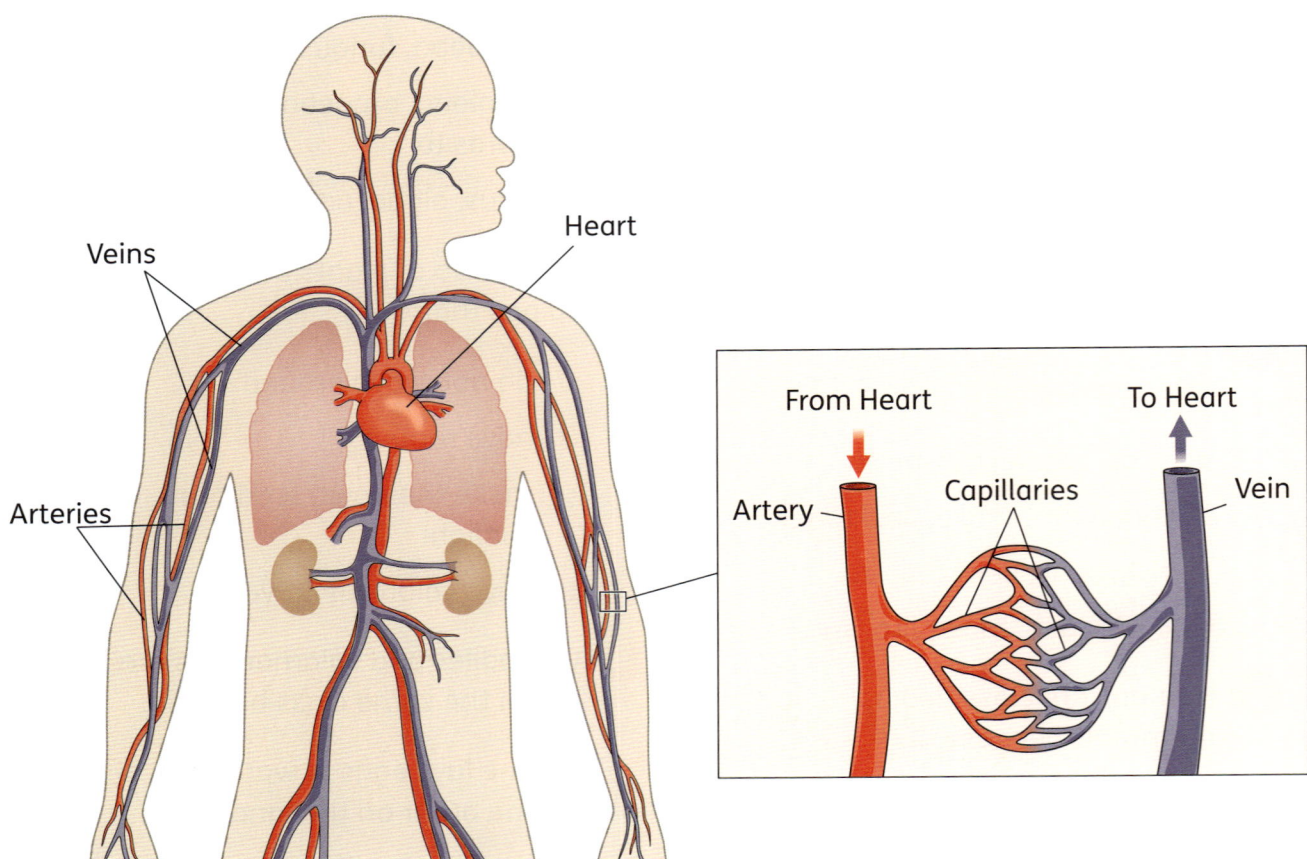

Veins

Heart

Arteries

From Heart

To Heart

Artery

Capillaries

Vein

The **heart** is a pump and it is connected to many tubes – the **blood vessels** (**arteries** and **veins**). The heart pumps blood out to the body along the arteries, and receives blood back from the body through the veins. These all make up the **circulatory system**.

Without the blood vessels the circulatory system would not work. The blood vessels are connected to a network of thinner and thinner tubes – the **capillaries** – that carry the blood to and from the millions of individual **cells** of the body.

You have learnt about cells of different types in Years 4 and 5 (for example egg cells, pollen cells, cells in the digestive system and the **respiratory system**). Your body has a great variety of cell types, each with their own special structure and **function**. They all need to be kept alive with food and oxygen whatever their function and wherever they are in the body.

Workbook page 10

The circulatory system is made up of:

• the heart
• the arteries
• the veins
• the capillaries
• the blood.

Even the blood is not all one thing – it too has several parts. It has liquid plasma, which is a yellow liquid that carries solid parts. There are three different kinds of solid parts in the plasma:

• the **red blood cells**
• the **white blood cells**
• the **platelets**.

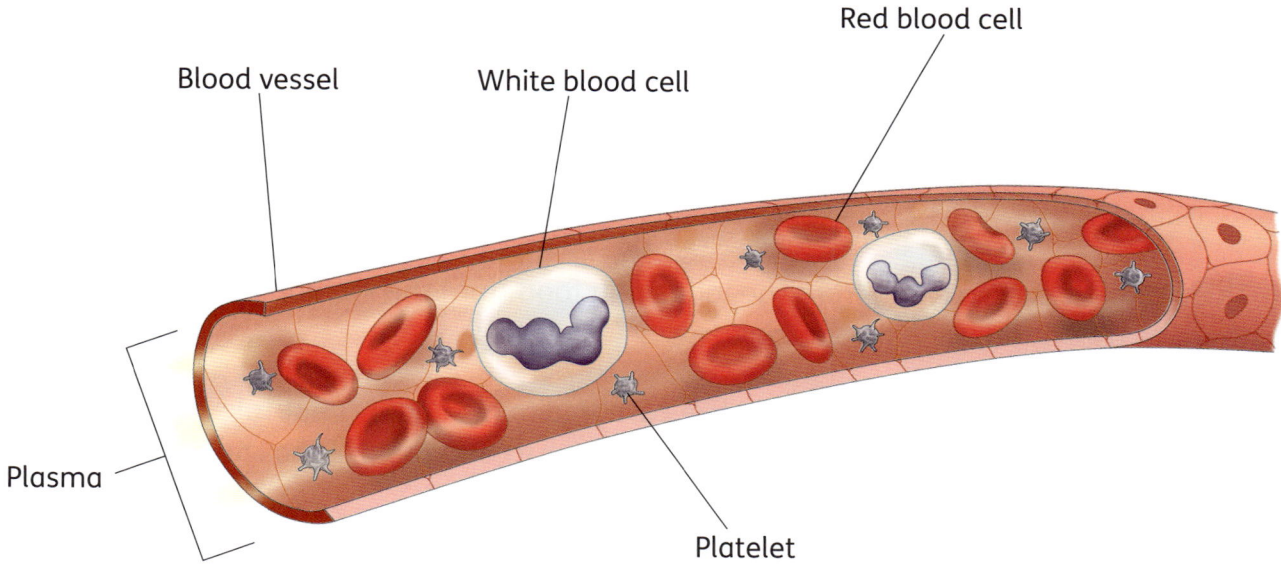

The blood carries **nutrients**, **hormones**, waste materials, oxygen and **carbon dioxide** to and from the cells in each part of the body.

• The blood supplies all the cells with oxygen and nutrients to keep them alive.
• The blood takes away carbon dioxide and other waste materials that the cells produce.

The blood connects the different parts of the body that need each other. It is a bit like a postal service!

Think of a **muscle** cell in your foot, for example.

- This muscle cell needs oxygen but it is far away from the **lungs**.
- The muscle cell needs food but it is far away from the **stomach** and **intestines**.

How can the muscle cell get oxygen and food? The blood will deliver it!

The blood picks up oxygen from the lungs and food from the intestines and delivers them to the muscle cell in your foot.

In return, the blood picks up wastes like carbon dioxide and **urea**. It takes the carbon dioxide back to the lungs (so you can breathe it out) and the urea to the bladder.

There is more water in the human body than any other substance. An adult's body is over 60% water! Water makes up a large part of all the major organs and even the skeleton.

The blood plasma is 90% water. Hormones and waste products are dissolved in the plasma. The blood also takes water and hormones to all parts of the body through the plasma.

The heart

The heart is essential to our survival because it pumps the blood around the body. Without the heart, the blood wouldn't move and so it couldn't take oxygen and nutrients to all the cells.

If a person's heart fails (stops), they might be able to have a **donor heart** or a mechanical heart put in its place. Both of these are amazing ways in which doctors can help to save people's lives! However, mechanical hearts are expensive and there is a shortage of donor hearts from people who have died.

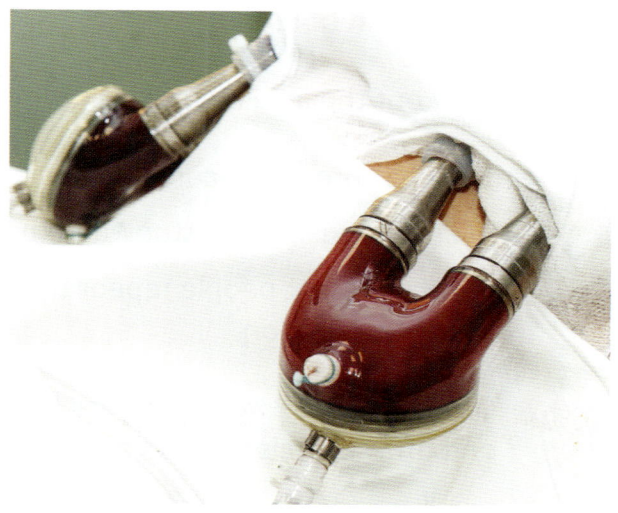

A mechanical heart

Major arteries and veins are also essential to our survival. If they are cut or burst, then we could die if doctors cannot mend them quickly.

Activity 1

You will need: a stopwatch, graph paper (or Workbook), ruler and a pencil.

1. Plan an **investigation** to find out how exercise affects the **rate** that your heart beats at.

   > Remember to make it a **fair test** by only changing one factor each time.

   a. You will need to **compare** the heart rate before and after exercise.

   b. Decide how long the person should exercise for, and the type of exercise they should do.

2. Draw a pair of axes on the graph paper.

   a. Discuss with your group what you should plot a graph of. What **variable** should you plot on each axis?

   b. What units should you use for the two variables on your graph?

3 **Predict** what the effect of exercise on the heart rate will be and write down your **prediction**.

4 Carry out the investigation.

    a Make sure you record the heart rate before and after exercise.

    b Practise using the stopwatch before beginning the test.

    c Let different people repeat the test until you have enough data to come to a conclusion.

5 On your graph, draw separate lines linking the heart rate before exercise and after exercise for each test.

    Write the initials of each person next to their line on the graph so you can see their change in heart rate.

6 In your groups, compare your results with your predictions. Come to a conclusion: 'What effects does exercise have on heart rate?'

7 Share your conclusion with the class. What else did you learn from these measurements? Try to explain what you have observed.

8 Now use your results to plan another investigation.

Choose either **a** or **b**.

    a You will change the type of exercise.

    b You will change the amount of time exercising.

9 Predict what effect this change will have on the person's heart rate, compared to their heart rate after the previous exercise.

✏️ Write down your prediction.

Prepare another graph paper before starting the measurements.

10 Draw another pair of axes on a new sheet of graph paper.

Before the exercise begins, measure each person's heart rate and record it on the graph.

11 Let each person try the new exercise.

   a Afterwards, measure their pulse rate and record it on the graph.

   b Draw lines connecting each person's pulse rate before and after exercise. Add initials to each one.

12 Look at the results and compare them with the first set of measurements.

   a Was your prediction correct?

💬 **b** Try to explain your results and share them with the class.

Exercise makes your muscles work harder than normal. This means that they need more **energy**.

Fats and sugars are dissolved in your blood. Your blood carries these to the muscles. Your muscles use the fats and sugars to make energy. Your muscles also need oxygen to release the energy from the fats and sugars. Your blood takes the oxygen to the muscles too. So, when you exercise, your heart has to pump faster to give the muscles the extra fat, sugar and oxygen they need.

Some forms of exercise are harder than others, as you may have discovered in your investigation. You know that walking is not as much effort as running and that you don't notice your heart beating faster when you walk.

Lifting weights, climbing, pulling heavy loads and cycling uphill are all harder forms of exercise and can make your heart beat much faster.

You also breathe faster when you exercise. The muscles need more oxygen, so the lungs have to put more of it into the bloodstream. The heart and the lungs work together to meet the muscles' needs during exercise. You know that you pant and 'run out of breath' after running hard, playing sport or running to catch the bus.

Transportation of water, nutrients and waste products

Humans take in all the water and nutrients we need through our mouths, in our food and drink.

As you learnt in Year 4, the mouth is the top end of the **digestive system** and solid food and liquids go into the mouth, down the gullet, into the stomach. **Digestion** takes place gradually as food moves through the digestive system. The nutrients are absorbed through the walls of the gut and into the blood at different points.

Water from our food and drink also passes out of the gut and into the blood.

The excretory system

Your body is kept alive by many processes, and these processes make waste products. If these wastes are not removed from your body, they can damage it.

The process of removing waste products from the body is called **excretion**. There are several organs that help with excretion. Each of these **organ** gets rid of a particular type of waste. Together, these organs form the **excretory system**.

Activity 2

You will need: paper (or Workbook) and a pen or pencil.

 1 Think of the things that you have to do every day to stay alive and well. Share your ideas with your group.

2 Of the things you discussed in question 1, choose which involve getting rid of waste products from the body.

a Discuss with your group which organs are used to get rid of each type of waste.

b Sort out the jumbled names of organs and waste products in the lists below.

c Match each organ with the waste **product** it removes and write down the pairs.

Organs:

nksi   siynked   glnsu   vielr

Wastes:

elbi   awtse   niuer   baocrn dodieix

3 Share your answers with the class.

The urinary system

This diagram shows one part of the excretory system of the human body. It is sometimes called the **urinary system**, because its function is to get rid of **urine**. Urine is mostly water, with several materials dissolved in it.

The main waste product is urea, a poisonous material produced by the **liver**. The liver also produces liquid called **bile**, a waste product that passes as a liquid into the intestine.

The blood carries the urea, and other poisonous materials, to a pair of organs that can remove them and clean the blood – the **kidneys**. These are located on either side of the body, at the level of our lowest ribs. In adults they are about 12 cm long.

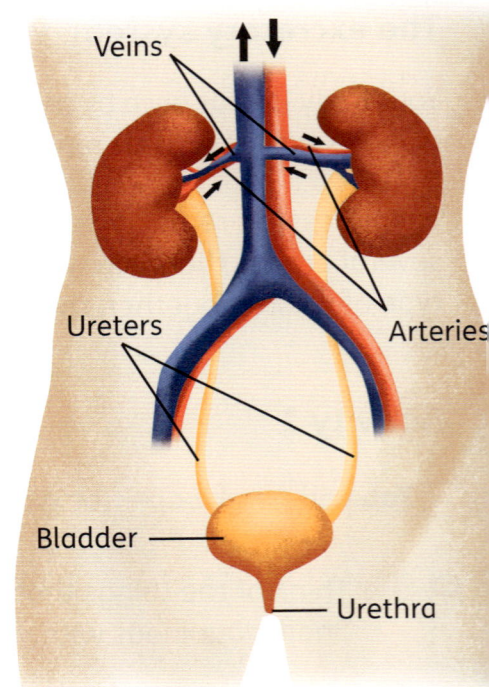

The kidneys remove waste products, along with a lot of water, as urine. The cleaned blood then returns to the heart through the veins to be re-circulated throughout the body.

The watery urine leaves the kidneys in tubes called the **ureters**, which carry it down to an 'elastic bag' of muscle called the **bladder**. The urine is stored in the bladder until the person feels that they need to empty the bladder by urinating. Babies do not have control of their bladders, but as we grow and develop, we learn how to keep the exit from the bladder closed, so that we can choose when to urinate.

The skin

Our bodies **sweat** all the time, but we notice it only when we have been very active, or when the weather is very hot. Then, the tiny drops of sweat join together on our **skin**. We see them and feel that our skin is wet. Sweating helps to cool us down and also excretes wastes from our bodies.

When do we sweat?

The skin is the largest organ in the body. Did you know that your skin is an organ?

The skin and the kidneys work together to get rid of some wastes from the blood – especially urea and some salts.

The surface of your skin is covered with tiny holes, called **pores**. Underneath each pore is a **sweat gland**. These excrete sweat, which is a mixture of water, urea and other waste materials.

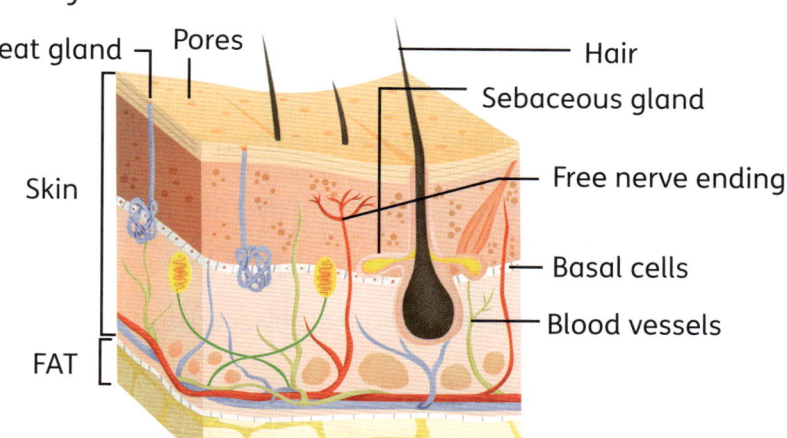

Elimination

In everyday speech, we often refer to **faeces** as a 'waste product' of our bodies. This is not really true – a waste product is something that our bodies make (like carbon dioxide or urea) when they take oxygen and food and turn them into energy.

Our bodies do not *make* faeces – they are mostly undigested food, which our bodies did not make. We ate this food at some point, and it just passed out again through the **anus**. Scientists call this 'elimination', rather than excretion, because the faeces has not been made by the body.

The liver

The liver is a very important organ. It lies between the digestive system and the circulatory system, so it acts like a doorway from one to the other.

As you learnt in Year 4, the liver has many functions.

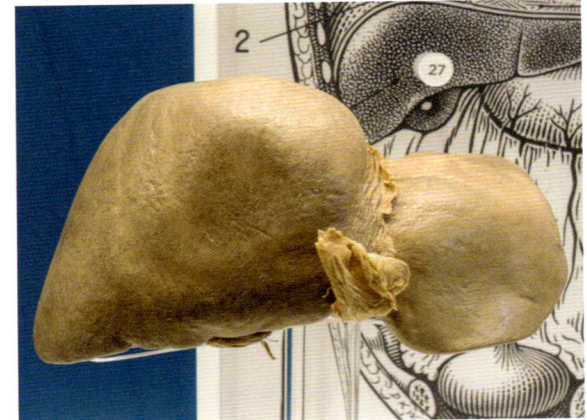

1 It converts and stores nutrients.

2 It produces bile.

3 It is vital to the body's other processes.

The liver's functions cannot be carried out by any other part of the body.

The liver converts nutrients from one form to another.

• Fats are converted into a form that the cells of the body can use.

• Some poisonous substances are converted into harmless ones.

• Waste products are turned into urea.

The urea, and the products from converting poisonous substances, are added to the blood and **excreted** from the body by the kidneys. They are waste products and the body cannot use them.

The liver also stores nutrients.

• It stores iron to make red blood cells.

• It stores vitamins A, D and B12.

The liver makes bile and proteins.

• Bile helps you to digest fats.

• **Proteins**, which are part of blood plasma, are made in the liver.

One other product that results from all these processes in the liver is heat. This heat helps keep our body temperature at the correct level. The blood carries the heat around from the liver to the rest of the body.

The final function of the liver is to keep the body's blood sugar, blood proteins, temperature, urea and bile all at the best levels for the health of the body. This is vital for the survival of all the body's organs and the processes that they carry out.

The liver is so important that if it fails a person will need a liver transplant to carry on living their life. If they cannot get a liver transplant they might die or they might need to go into hospital for a treatment called **dialysis**.

Transportation of water, nutrients and waste products in other animals

You have learnt that our blood carries these things around our bodies:

• water

• nutrients (digested food)

• waste products.

These three things happen in most kinds of animal too. Mammals, birds, reptiles, amphibians and fish all take food into their bodies through their mouths and excrete water, one way or another.

bird

lizard

frog

fish

bat

Water is part of most foods, but land-based animals generally need to drink water as well. Most animals that live in lakes, rivers or the sea don't need to drink. They get enough water from their food.

There are many different kinds of invertebrates, but to remain alive they all need to take water and nutrients into their bodies. Many have a mouth and digestive system, as in vertebrate animals. Many also have circulatory systems that transport and distribute the water, nutrients and waste materials around the body.

earthworm

crab

locust

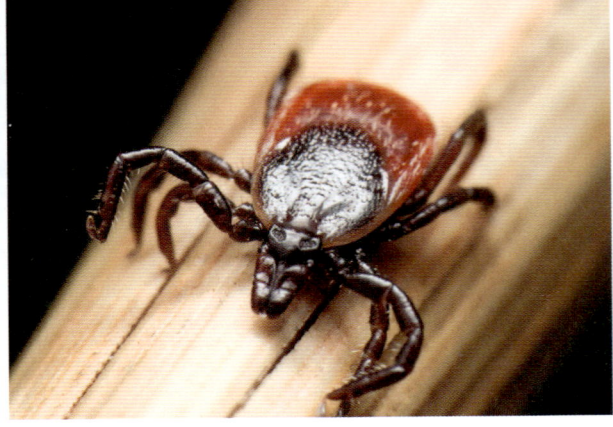
tick

Invertebrate animals

Protozoa and bacteria don't have either a digestive system or a circulatory system, but they still need things like nutrients and water. Because they are only a single cell, the cell wall allows materials to pass through into the inside of the cell.

Diet and exercise

In levels 2 and 3 you learnt some important things about **diet** and health.

1 There are five main food groups:

fats and **oils**   **fruits** and **vegetables**   **vitamins** and **minerals**

proteins: **animal proteins** and **plant proteins**   **carbohydrates**

Can you match each food in pictures (a)–(p) with one of the groups?

(a) sliced bread   (b) peas in pod   (c) butter   (d) bowl of rice

(e) slice of cheese   (f) carrot   (g) eggs   (h) chicken leg

(i) broad beans in pod   (j) bottle of orange juice   (k) sweet patatoes   (l) white patatoes

(m) sliced mango   (n) fish   (o) banana partly peeled

2 To have a healthy diet, you must eat a mix of foods from each of these groups. This is called a **balanced diet**.

3 It is important to get regular **exercise** to stay healthy. You exercise when you make your body work hard, use your muscles, bend your joints, breathe deeply and make your heart work harder and faster.

Exercise helps to burn off extra fat if we have eaten too much.

4 It is important to rest and sleep well to stay healthy.

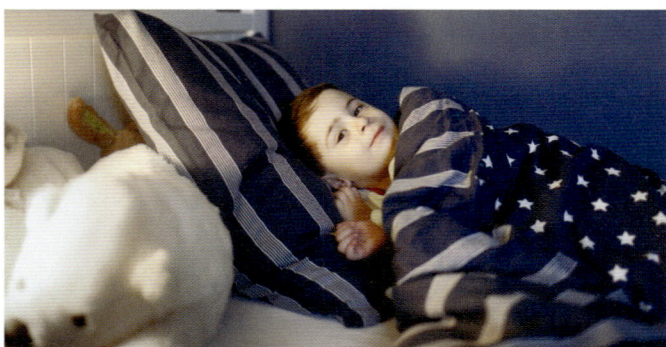

In the same way that you need a balanced diet to remain fit and healthy, you need a good balance between exercise and rest.

5 Personal hygiene (staying clean) is important for you to stay healthy. You should:

- clean your teeth
- change your clothes regularly
- wash your hands, especially after using the toilet
- wash and comb your hair.

These simple habits help to reduce the germs on our bodies and this helps to protect us from **infections** and illness.

Activity 3

You will need: paper (or Workbook) and a pen or pencil, secondary sources of information.

1 Discuss the five different food types in your group.

    a Research why each of the food types is important for your body.

    b Agree examples of two foods from each food type that people in your group have eaten recently.

2 Present your findings in a table:

Food type	Importance for our bodies	Two examples eaten

3 a Share your research with the class.

    b Add to your table any extra examples given by your classmates.

Activity 4

You will need: paper (or Workbook), graph paper, a ruler and a pen or pencil, coloured pencils or paints.

1 Plan with your group how to find out what types of exercise your classmates do each week.

   a Discuss with your group what questions you will ask. This is called a survey.

   b Decide how to record your results.

   c Write your plan and show it to a teacher. When your teacher is happy with it, carry out your survey.

2 **a** Ask each of your group to tell you the one exercise they carry out the most in a week.

   b Record your findings in a table like this:

Person	Exercise

3 **a** Decide in your group which is the best type of chart to present the data.

   b Draw a chart to show your results.

   c Make sure you label the chart correctly.

4 **a** Share your chart with the class.

   b Conclude what is the most popular type of exercise is in your class. Write down your conclusion.

5 **a** What other questions could you ask your classmates about their exercise to add useful data to your survey? Hint: you could ask how long they exercise for or how exercise makes them feel.

   b Write down two ideas.

6 **a** Ask members of your group the two extra questions and record your findings in extra columns in your table.

   b Share your findings with the class.

7 Use your data to design a colourful poster or web page to share the class's findings about exercise.

Drugs

A **drug** is any substance, other than food, that causes changes to our body.

People can take drugs by:

• swallowing them
• breathing them in
• injecting them into their bodies
• applying them in some other way.

Some drugs are good and helpful to our bodies. Other drugs are bad and cause harm.

There are three groups of drugs:

• **prescription** drugs
• over-the-counter drugs
• **illegal** drugs.

Drugs that we buy in **pharmacies** and **chemists** all have very important information on their packaging. This information is to protect us from harming ourselves by taking the drugs in the wrong way, or by taking too much of the drug.

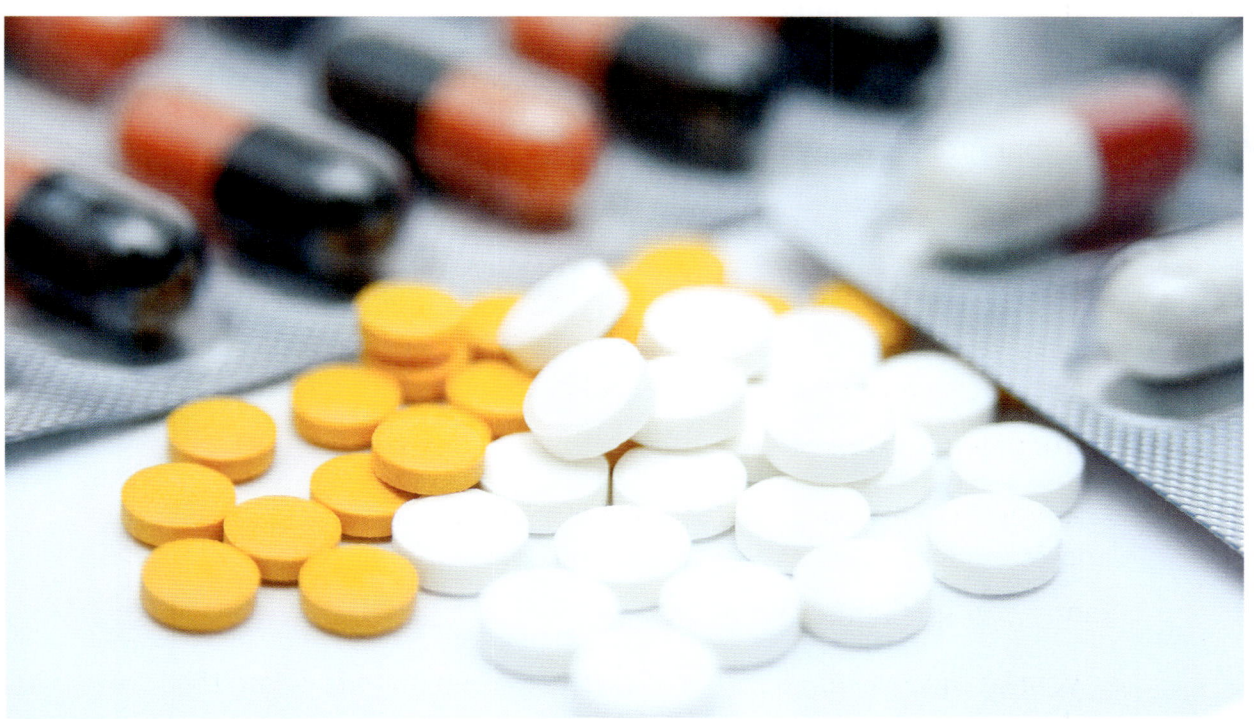

Any drug can be dangerous if we take it in the wrong way, or take too much of it. This is true of prescription drugs and over-the-counter drugs, as well as **prohibited** drugs.

To use prescription and over-the-counter drugs safely we must follow the instructions about the **dose** (how much to take, and how often), what age we have to be to take it, and the warnings about any side effects.

Taking more of a drug does not mean we will gain more benefit from it. It might lead to serious damage to our bodies, or even death. For example, organs such as the liver and the kidneys can be damaged by high drug doses and, if the damage is very bad, it might not be possible for them to recover.

Children are in the greatest danger from drugs, because their bodies are smaller and so they are more easily damaged by an overdose of drugs. This is why drugs must be stored safely at home, in a place where young children cannot reach them.

Activity 5

You will need: paper (or Workbook) and a pen or pencil.

1 Discuss with your group:

    a *What are the benefits of drugs?*

    b *What are the harmful effects of drugs?*

2 Make notes of your group's answers to these two questions.

3 Share your group's answers with the class.

The benefits of some drugs

Prescription and over-the-counter drugs have many benefits.

Some drugs, like antibiotics, can cure diseases by killing the organisms that have invaded our bodies.

Other drugs, like paracetamol or throat lozenges, can take away symptoms, such as the aches and sore throat of a flu infection, without killing the virus causing the flu.

Some drugs can prevent us being infected. They can give us protection against disease. Anti-malarial drugs are examples of this kind; they help protect us from getting malaria.

If a person's body is not processing sugar in the right way, they might need treatment for **diabetes**.

A person can inject a drug called insulin into their arm or side. This helps the body to remove the sugar from the blood.

This type of drug has saved many lives!

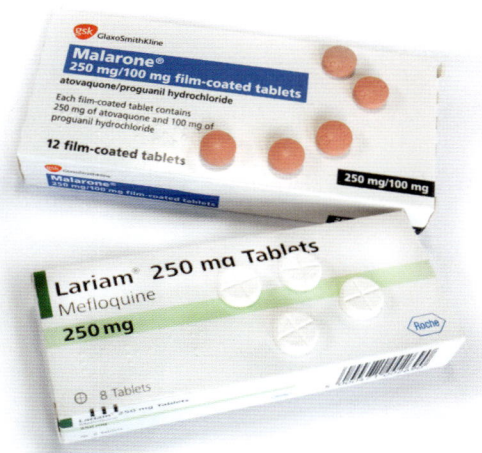

The harmful effects of drugs

Drug abuse is when a person takes a drug when it is not for medicine, or if they use too much of it. Drug abuse does not only happen with illegal drugs, but can happen with prescription and over-the counter drugs too. They can all have harmful effects on the body.

The major problem with using all these substances is that their use can easily become **addictive**. This is why most of them are illegal drugs.

Alcoholic drinks are legal in some countries, but they do contain the drug alcohol.

Consuming alcohol in large quantities over long periods can:

• destroy cells in the liver, which can lead to a serious illness called cirrhosis that can kill a person

• damage the lining of the stomach and cause painful ulcers.

Drinking too much alcohol also makes a person have less control over their body or behaviour. People should never drive a car if they have drunk more than the legal amount of alcohol.

Tobacco, which is generally smoked, also damages the body.

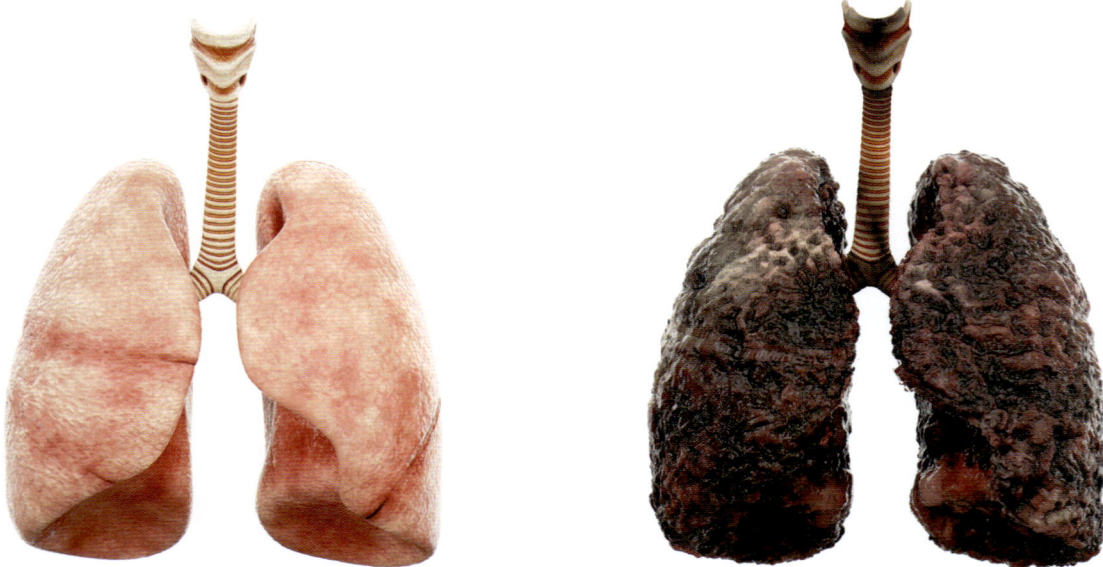

Healthy lungs alongside lungs damaged by smoking

- The tiny air spaces in the lungs become filled with tar from the smoke

- This tar can lead to breathing problems, or cancer of the lungs. Lung cancer is a major cause of death among smokers (7 out of 10 cases of lung cancer in the UK are caused by smoking).

- Smoking increases the risk of developing more than 50 serious health problems, such as heart disease, stroke and cancer in other organs.

The **nicotine** in cigarettes means that they are addictive, and people often need help to stop smoking.

Other addictive drugs, such as marijuana, cocaine, heroin, barbiturates (sleeping pills) and amphetamines (which stimulate the body and speed things up) have many harmful effects on the body.

Warnings to smokers on cigarette packets

Science in Action

Since the 1960s doctors have been able to transplant the heart from someone who recently died to a person whose heart is failing or damaged. The transplanted heart can give the chance of a healthier life and extra years to the person who receives it. Thousands of these operations take place across the world every year.

Most of these operations are very successful, but because this involves major surgery there can be risks involved. The actual replacement of a heart can be seen as a simple 'plumbing job' – cutting and connecting tubes – but the effects of this process have to be carefully managed.

The 'new' heart is recognised as 'foreign' by the body it has been transplanted into, and so it can be 'rejected' by the body's defences. Doctors have to use drugs to prevent this, but the drugs can increase the risk of catching infections.

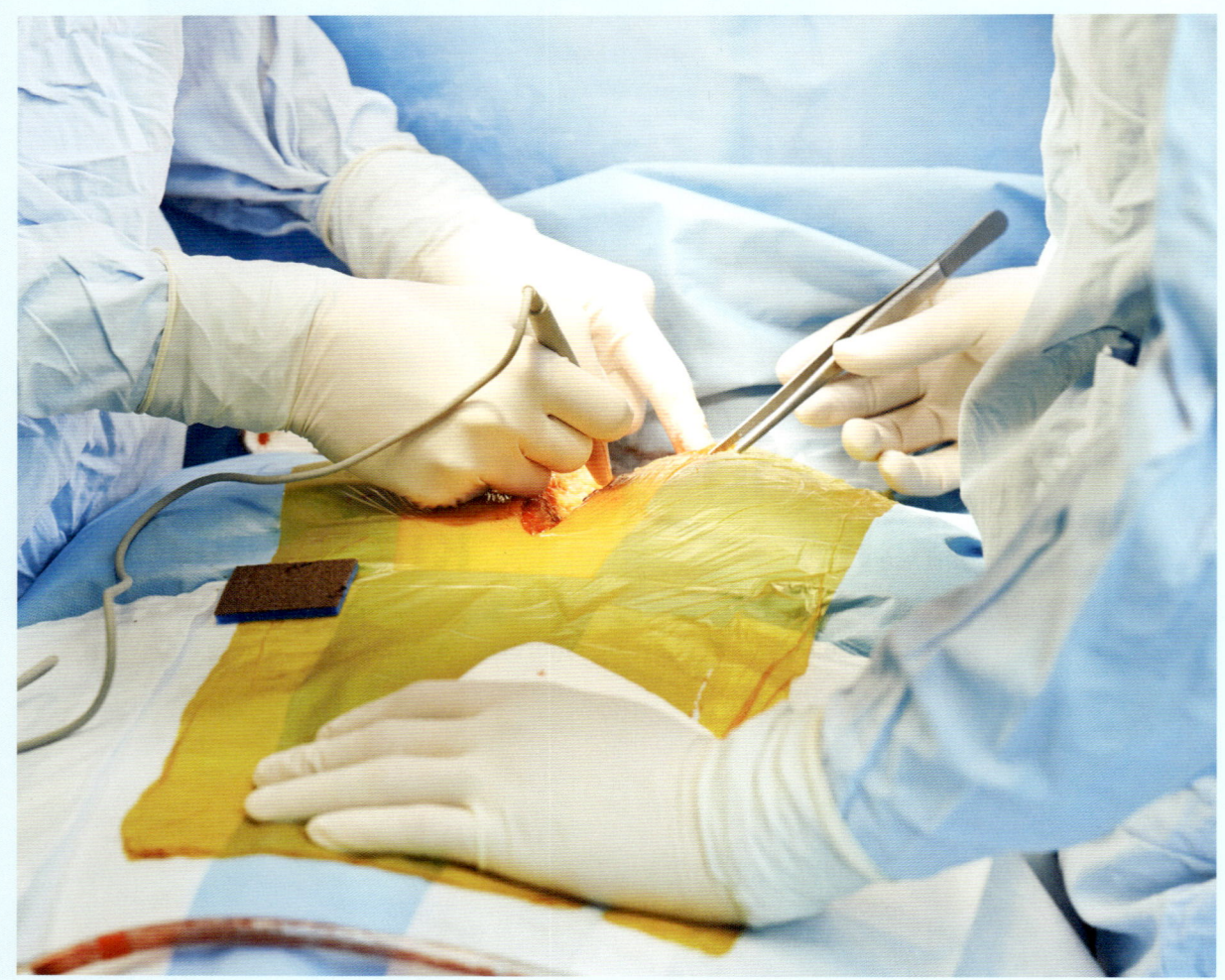

Doctors performing heart transplant surgery

End of unit check

Match the words in Box A with words in Box B. Write down the pairs of words.

BOX A
excretion   exercise   heart   diet
blood   water   drugs   heartbeat
artery   urea

BOX B
tube   pump   balanced   solvent
pulse   harmful   urine   liquid
health   kidneys

_____   _____

_____   _____

_____   _____

_____   _____

_____   _____

UNIT 3
Variation and the theory of evolution

Fossils

Fossils and rock layers

In Year 3 you learnt that some rocks contain **fossils** of living things from long ago.

Here are some examples.

   a plant and a fossil

   a fish and a fossil

   a trilobite and a fossil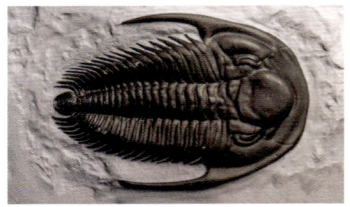

   an archaeopteryx and a fossil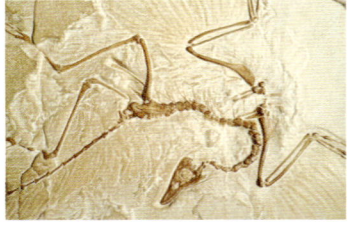

   a dinosaur and a fossil

Fossils are like little 'cloudy windows' that give us just a glimpse of what some life on Earth was like a long time ago. What we find as fossils is only a tiny sample of the many, many creatures that have existed on Earth over the years since life first appeared. Most have left no trace.

Bones and shells are most easily **fossilised**. Soft parts of animals and plants are not often found as fossils. These parts rot away when the animal or plant dies.

Sometimes the space where the animal's body was buried in the mud gets filled by minerals and so we have a copy of the creature. The mud acts like a mould.

The fossils shown opposite are examples of animals and plants that are no longer found living anywhere on Earth. They have all become **extinct**.

As you learnt in Year 3, the rocks of the Earth's **crust** are in layers, with the oldest rocks generally lower down than the newer ones. This helps **palaeontologists** to estimate how long ago the fossilised animals or plants lived.

Looking at fossils in the deeper layers of rock helps us understand more about the earliest forms of life. Fossils in the higher layers of rock help us understand more about later forms of life.

Each layer of rock in this table represents a period of time in history. The deepest layer – the Precambrian – is the oldest. The highest layer – the Quaternary – is the newest. We are living in the Quaternary period.

Some of the fossils that were found in each layer are shown in the table.

Period	Plant	Animal
Quaternary	Many modern trees	Woolly mammoth Modern humans
Tertiary	Beech tree	First Primates
Cretaceous	Cycad	Tyrannosaurus
Jurassic	Gingko tree	First mammals
Triassic	Conifers	First dinosaurs
Permian	Seed ferns	Dimetrodon
Carboniferous	Giant horsetails	Early reptiles
Devonian	Plants with seeds	First amphibians
Silurian	Land plants	First bony fish
Ordovician	Sea weeds	First vertebrates
Cambrian	Simple algae	Trilobite
Precambrian	None	None

Activity 1

You will need: paper (or Workbook) and a pen or pencil.

1. Look at the table and follow either the development of the plants or the animals from the bottom to the top. Write down what the table shows you.

2. Now do the same with the other column.

3. Compare what you wrote down about how plants and animals developed. Come to a conclusion and make a generalised statement about how living things have changed over time.

4. Share your ideas with the class and explain them.

5. *Did humans ever run away from dinosaurs like Tyrranosaurus Rex? Which came first – plants on the land or animals on the land? Why?*

    Discuss these questions with your group. Make up other questions to ask one another, using the information in the table.

The oldest rocks have no fossils at all. This is true all over the Earth. From this, scientists have suggested that there was no life of any kind on Earth at the beginning.

The earliest fossils are dated as between 3000 and 4000 million years ago. These bacteria-like fossil organisms have been found in several sites across the world.

Because there are new fossils in each new layer of rock, scientists think that new plants and animals have lived on the Earth as time went by. Some of the organisms we find as fossils are now extinct. Many other plants and animals have taken their places.

Adaptation

Your school is surrounded by **habitats**. Some are very small and others are enormous. A building itself creates habitats for certain animals. Trees, especially when they are old and large, also provide habitats for some animals and even other plants. Each **environment** has many, many different habitats for its animals and plants.

Activity 2

You will need: two different habitats, paper (or Workbook) and a pen or pencil.

1 Discuss with your group which two habitats you will **investigate** in your local area. Try to choose two that are very different.

2 Plan what you will look for and how you will record what you observe.

> Remember that drawing, measuring, counting and writing notes are all good ways to collect and record data.

3 Go outside with your notebook and pen and investigate your chosen habitats.

a Collect enough information to be able to give a clear and full description of the animals and their habitats to the class.

b Try to describe what each habitat is like. Use these questions, for example.

- Is it hot and dry?

- Is it shady?

- Is it wet?

- Is it bare soil, rock or sand?

- Is it in or near water?

- Are there plants, and are they close together or spread out?

4 Return to the class.

    a Discuss with your group the information you have collected.

    b Decide how you will present your descriptions to the class.

5 Present your description of each habitat to the class. Try to answer questions about the habitat from others in the class.

Look at these animals.

Choose one of the animals.

Discuss what the habitat of your chosen animal is like.

- What is special about the place it lives in?

- What is special about the colour of the animal?

- How might this help the animal?

Describe your answers to the class.

There are many different natural habitats on Earth.

Some habitats are so cold that very few things can live there, for example the Arctic and the Antarctic.

Others are so hot and dry that very few plants or animals can live there, for example deserts in South America, Africa and Asia.

Very few people live in these extremely cold and extremely hot places. It is too difficult to stay alive.

Plants and animals can be found in most places on Earth. Each place has its special animals and plants that 'fit' the conditions. Usually the amounts of water, light and heat are the most important features of any habitat.

Look at the animals in the pictures and discuss what is wrong with the habitats they are in.

Tell the class what you think.

Now look at the animals in the pictures below. Each of these animals is shown in its natural habitat

Use these words to complete the sentences below:

tadpoles	tiger	hide	feed	warm	cold
frogs	slide	smooth	water	flowers	
snakes	eggs	birds	penguins	sea	shape
fish	nest	bees	pond	deserts	

1 The _____ lives in the jungles of India. Its stripes help it to _____ when it is hunting.

2 _____ live in large numbers in the _____. They swim and _____ together. Their _____ helps them to move easily through the _____.

3 The Antarctic is a very _____ habitat so the _____ and other animals have to have a way of keeping _____. The feathers protect the _____ from the cold.

4 Some _____ live in very hot, dry places such as _____. Their skin is dry and it is very _____, which helps them to _____ over the sand and rocks.

5 _____ and other insects visit _____ to collect food. They must use it for themselves or take it back to their _____ to feed their young.

6 A _____ is a good place for _____ because they must have fresh water to lay their _____ in. Adults can come out of the water, but the eggs and _____ must live in it as they grow.

These animals all 'fit well' in their habitats. Some scientists call this idea **adaptation**, which says that the features of an animal are adapted to help it survive in the habitat it lives in.

Any changes that offspring inherit from their parents could make them better adapted and so more likely to survive in a particular place. You will learn about this in the next section.

Variation in offspring and the theory of evolution

Activity 3

You will need: paper (or Workbook) and a pen or pencil.

1. Look at the cat family and the human family.

   a Look for differences between parents and their offspring in each picture.

   b Write down the differences you see.

Parents and offspring

2. Can you think of any other differences there might be between parents and their offspring that you cannot see? Write these down too.

3. Think about what you have learnt about **reproduction** of humans and other animals. Use this knowledge to explain why there are differences between the parents and the offspring in the pictures.

4. Share your ideas with your group and listen to those of others.

If the parents and offspring had all been identical, then you would have been surprised!

You know from your own family that children are never exact copies of their parents. You may have some **characteristics** of one of your parents – such as hair colour – and some of the other parent – such as left-handedness. This is called **inheritance**.

Each of us is a mixture of our parents. This is true of all living things. The pollen grain and the ovum combine to produce a new plant that may only vary a little from its parents, but it could vary a lot. The egg and the sperm from the cat parents will not produce an exact copy of either.

This is called **variation**. Each generation is different in some way from its parents. Over one or two generations the changes will be small and we may not notice them, but over thousands of generations there will be bigger changes.

The theory of evolution

Charles Darwin (1809–1882) and Alfred Russel Wallace (1823–1913) were biologists. They suggested a theory (an idea) to explain how:

- the plants and animals we see on the Earth now have not always existed
- many plants and animals that did exist have disappeared.

This has become known as the theory of **evolution**.

The theory says the following.

- Some changes that living things inherit from their parents are good as they help the organism to be better adapted to live in its environment. They give the individual organism an advantage of some kind.

- All life can be **competitive** and so, for example, if a plant inherits the ability to grow more quickly than other plants, it may capture more sunlight, produce more food, have more flowers and seeds and leave behind more offspring. Weaker plants may lose out and leave fewer offspring. Over a long period, the faster-growing plants may totally dominate a habitat and the weaker **species** dies out in that particular location.

- The same thing might also apply to animals.

  - If a predator inherits the ability to run faster than other predators, it will catch more **prey** and so be more likely to survive in its habitat.

  - A bird that inherits a longer beak which can go deep into flowers has more to eat, and so it will be more likely to survive in its habitat.

- In this way, the disadvantaged individuals and species might gradually fade out and the advantaged individuals and species might increase in number.

Changes in the environment

Sometimes the living things in a habitat can change because the habitat itself changes and the animals that lived there are no longer adapted to the new environment.

In the **ice ages**, it was too cold for many species to survive. They became extinct. Other creatures, that were better adapted to the cold, took their place.

An extinct elephant: the woolly mammoth

The **woolly mammoths** were perfectly adapted to the freezing temperatures of the Arctic regions of the northern hemisphere, in Europe, Asia and North America. Their blood contained something that stopped it freezing.

However, when the climate warmed up, their frozen habitat disappeared. At that time, they were also prey for human hunters. Because of these two things, eventually woolly mammoths became extinct.

Humans now live in almost all the land-based habitats on Earth. It is amazing how we can survive both the hot temperatures of the tropics and (with the right equipment) the freezing temperatures of the Arctic and Antarctic regions.

Use the words from the box to complete the sentences.

```
    change   generation   extinct   slowly   offspring   survival   vast

  evolution   fossilised   fur   variation   humans   complex   parents

compete   freeze   advantages   first   tiny   palaeontologists   extinct

        single-celled   human   mammoth   habitat   millions

adapted   observation   offspring   process   habitats   species   fossil
```

1. Only a _____ number of creatures that have lived on Earth become _____ and can be found by _____.

2. _____ numbers of animals and plants have become _____ and this process is still continuing each year, often caused by _____ activity.

3. Some scientists believe that the _____ living things were very simple _____ organisms and the development into the _____ variety of life has taken place very _____ over _____ of years.

4. The theory of _____ is based on the _____ records and _____ of plants and animals in many _____ across the world.

5. _____ never exactly match their _____ and this _____ is the mechanism that drives the _____ of evolution.

6. Small changes in each _____ can give _____ to a species that help it to _____ with other _____ in the habitat.

7. The organisms that are best _____ to a particular _____ have the best chance of _____ and the best chance of having _____ .

8. The woolly _____ is one example of a well-adapted _____ animal, with thick _____ and blood that did not _____ , but it became _____ because of climate _____ and being hunted by_____ .

Vision

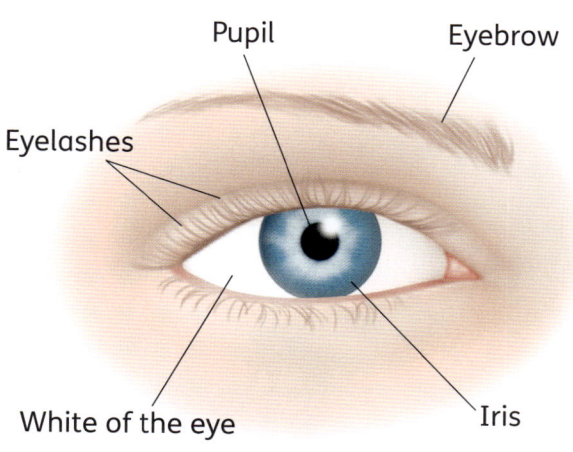

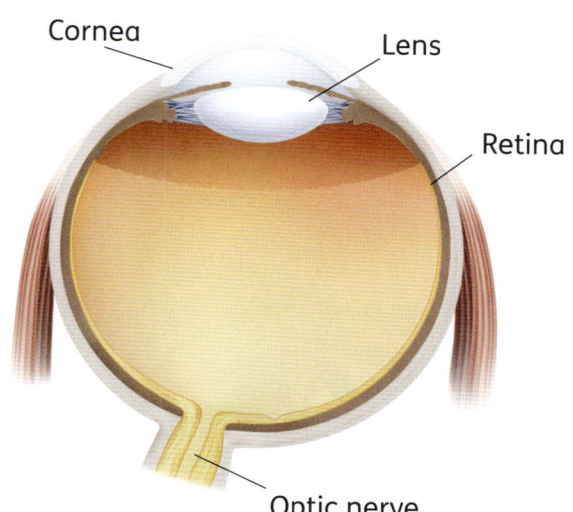

Human eyesight

The front of the eye is covered with a transparent **cornea**, which bulges out slightly.

The cornea protects the more delicate parts of the eye, which are behind it. Light passes through the cornea and enters the eye through the **pupil**.

The pupil is a hole at the front of the eyeball. It appears as the black, circular centre of the eye. The size of this hole changes, depending on the amount (intensity) of light shining on the eye.

The iris is the coloured ring of tissue that surrounds the pupil. Tiny muscles in the iris change the size of the pupil.

• When a lot of light shines on the eye, the iris automatically relaxes. This closes down the size of the pupil to a minimum.

➡ *Workbook page 33*

- When very little light shines on the eye (like at night), the iris automatically contracts and opens the pupil to its maximum size. This is 16 times bigger than its minimum size!

We do not have to think about the size of our pupils. The iris automatically changes the size of our pupils to match the amount of light.

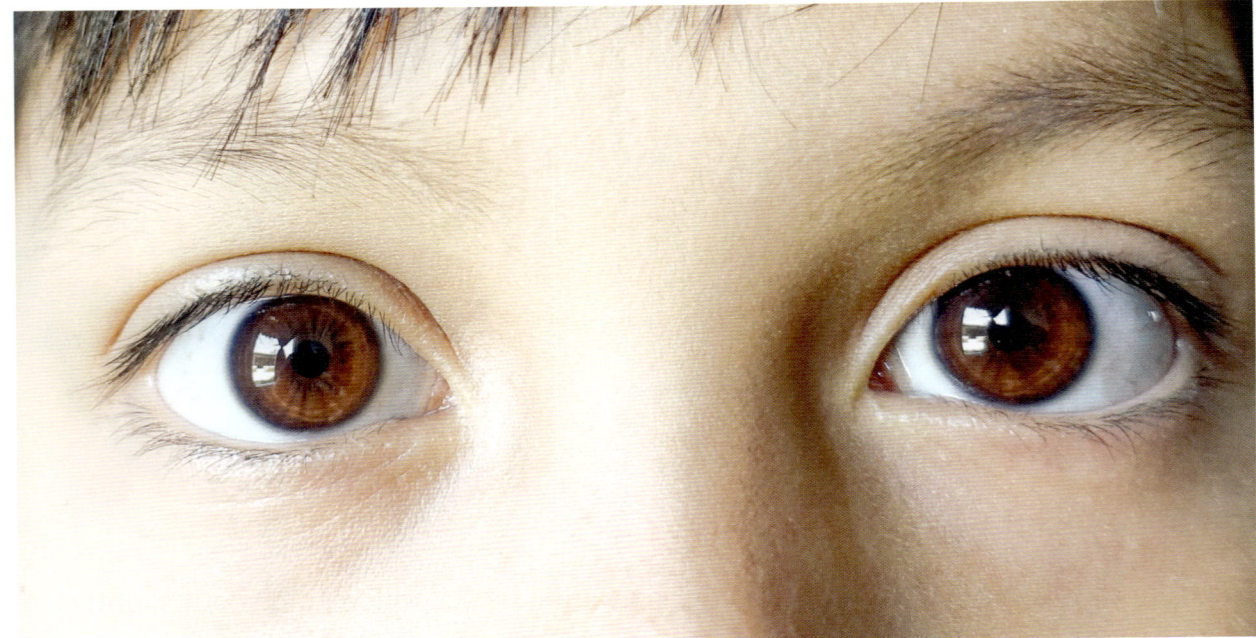

What wonderful things our eyes are!

Light travels from a source to our eyes

All light sources send out **rays** (or beams) of light.

Think of sunbeams and moonbeams. Think of the beam of light from a torch or a car headlight. Light from some sources travels in all directions – for example light from candle, the Moon or the ceiling light in a room. Light from other sources is directed into a beam that points in a particular direction – for example light from a torch, a television or a computer screen.

We can see a light source when we look at it. This is because light rays from the source pass through the transparent cornea and enter our eye.

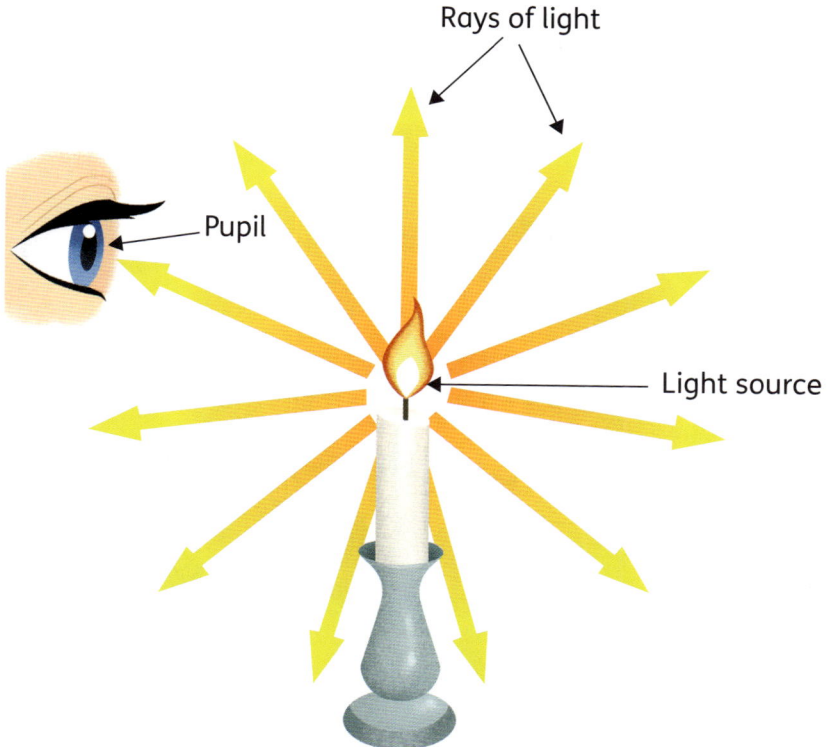

Rays of light

Pupil

Light source

We see a light source when light rays from the source enters our eyes.

Light travels in straight lines

Activity 1

You will need: a drinking straw or other flexible tube (like part of a garden hose), a light source (such as a candle or torch), paper (or Workbook) and a pen or pencil.

1 Hold the tube so that it is straight. Look through the tube at the light source. Record what you observe.

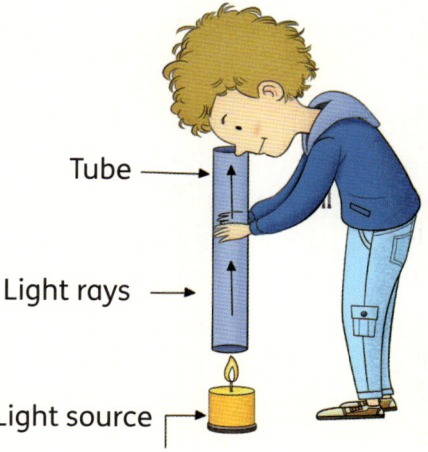

Tube

Light rays

Light source

🔍 **2** Bend the tube slightly so that it is not completely straight.

Then look through the bent tube at the light source. Record what you observe this time.

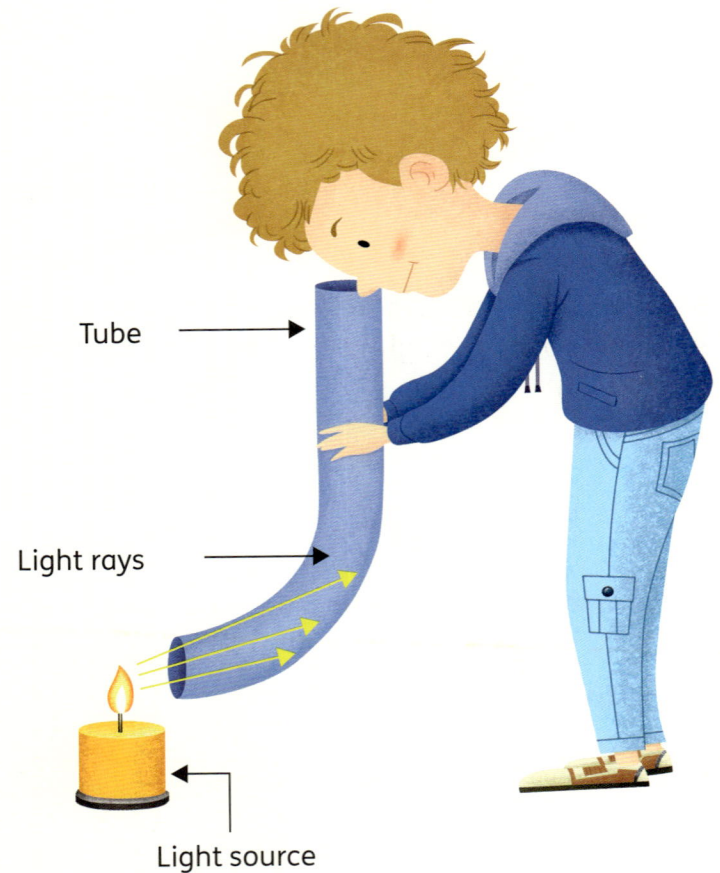

Tube

Light rays

Light source

💬 **3** Discuss the result with your group and try to explain it.

💬 **4** Share the group's results and explanation with the class.

Light travels in straight lines from its source. This means that if anything **opaque** is in the path of the light, it will block the light. The light cannot 'go around a bend' and carry on its way. In a straight tube such as a drinking straw, the light rays travel along it without being blocked. That is why we can see through the straight tube.

If the tube is bent or curved, there is no straight pathway for the light and so it cannot travel through the tube. We cannot see the object through the tube because the light is blocked by the bend or curve.

Activity 2

Because light travels in straight lines, we can create images on a flat surface by reflecting light.

> **You will need:** two shiny objects, a mirror, a torch, a wall, paper (or Workbook) and a pen or pencil.

1. Use the mirror to create as many different effects as you can.

2. Switch on the torch. Using a mirror, can you make the light from your torch beam shine on a wall?

3. When you have succeeded, draw a diagram of how you did it. Use arrows to show the direction and path of the light from its source to the mirror, and from the mirror to the wall.

4. Collect two shiny objects and take them outside into the sunlight.

5. Try using the shiny objects to re-direct the sunlight onto a wall or other shaded surface. How did you do it?

6. Back in class, draw a diagram of what you did with the shiny objects.

 **7** Discuss your results with your group.

 a Try to explain how the light was re-directed by the mirror and the shiny objects.

 b Compare the different amounts of light you could shine on the wall with the two shiny objects. What is similar about both objects?

 **8** Share your results and explanations with the class.

Shiny objects, including mirrors, **reflect** light. The light hits the surface and 'bounces off'. A mirror is designed to produce a perfect reflection.

Many metal and plastic objects reflect light well.

Most mirrors are glass with a thin layer of **aluminium** on the back, which reflects light very well.

Activity 3

You will need: a mirror, paper (or Workbook) and a pen or pencil.

1 Write out the alphabet in capital letters on a sheet of paper. Make each letter at least 2 cm tall.

2 Hold up the paper in front of the mirror. If the mirror is large enough, you will see all the letters at once. If it is small, look at each letter one at a time.

 a What do you notice about the letters when you look at their images in the mirror?

 b Write down what you see.

 c What do you notice about the images of these letters?

 d Write down what you see.

3 Try to make a word that looks exactly the same on the paper and in the mirror.

4 Look at yourself in the mirror and touch your left ear with your right hand.

5 Discuss the results of all these investigations with your group.

   a Try to come to a conclusion about how mirrors make reflections.

   b Share your words from question 3 and your conclusion with the class.

How we use light rays

Now you know that light travels in straight lines, you can understand how we see things.

Light either comes directly to our eyes from a light source (for example a lamp), *or* it is reflected off the surface of objects, and travels to our eyes.

Opaque objects that lie between a light source and our eyes block the path of light to our eyes. This is why we cannot see objects around a corner or behind a wall. We say they are 'out of sight'. Light has to have an open pathway from an object to our eyes for us to see things.

Periscopes are used in submarines to help sailors spot ships on the surface. They use the fact that light travels in straight lines.

A child using a periscope

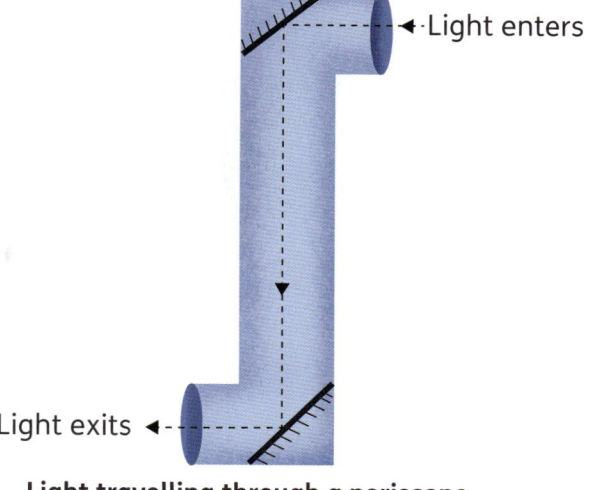

Light enters

Light exits

Light travelling through a periscope

Activity 4

You will need: cardboard to make a box for the periscope, two identical small mirrors, sticky tape, scissors, ruler, paper (or Workbook) and a pen or pencil.

1. Your teacher will help you to make a periscope from cardboard.

2. Discuss with your group what you will do with the two mirrors. The drawing of the periscope on page 59 should give you a clue.

> Remember that light travels in straight lines, through the top window, down the tube and out of the bottom window.

3. Fix the mirrors in the right places.

4. Test the periscope to see if you can see out. If not, change the position or angle of the mirrors and try again.

5. Draw a diagram of what the light is doing as it travels from the object you are looking at, inside the tube and out into your eyes.

6. Go outside to try out your periscope. Look over walls or round corners. If you are not able to go outside, use it in the classroom to see things that are normally hidden.

7. Share your periscope and diagram with the class. Explain how your periscope works, using what you know about light travelling in straight lines.

View from a periscope on a submarine

Activity 5

🔍 **1** Go outside on a sunny day and look at shadows. Notice their size and direction.

If you are not able to go outside to do the next activity, stand near a window with direct sunlight shining on you.

🔍 **2** Look at your own shadow and change it to create different shapes. Make at least three different shadow shapes.

3 Show the class the shapes you make. Look at the shapes that other people make.

4 Try to 'lose' your shadow – make it disappear somehow.

When you have done this, tell the class how you did it. Observe how other people did it. Were there different ways to make your shadow disappear?

5 Back in class, discuss with your group what you have observed outside.

   a Try to answer the question: how are shadows formed?

   b Share your group's ideas with the class.

In Activity 5 you thought about how shadows are formed. In Activity 6 you will test those ideas to see if you were correct.

Activity 6

You will need: a camera, a torch, a ruler, a few objects to make shadows with, paper (or Workbook) and a pen or pencil.

1 Plan how you can test your ideas about how shadows are formed. You should use a torch and a number of objects to make shadows with.

   a Make it a fair test. Think about what conditions of the test must be made the same for each object.

   b Decide what observations you will make and how you will record them. You can use a camera to help you with this.

   c Write your plan down and show it to your teacher.

2 Carry out your test and record what happens.

3 Look at your results about the shadows that different objects made.

   a Compare the shadows.

   b Try to write down a conclusion about how shadows are formed. Use the results of your experiment to help you.

4 Share your results and conclusion with the class. Listen to other people's conclusions. Did you all think the same thing?

To form a shadow you need:

• a source of light (a torch, the Sun or something else)

• an object that does not allow the light to pass through it completely (it must be either translucent or opaque).

If you have both these things, the shadow of the object forms on the opposite side from the light source.

Shadow shapes

You will have seen that the shapes of shadows are similar to the shapes of the objects making the shadows.

Straight-edged objects did not make curving shadows. Curved objects did not make straight-edged shadows. Think about why this happens, using your knowledge of how light travels. Tell the class what you think.

Activity 7

You will need: three objects of different shapes, blank paper, coloured pencils, a light source.

1. Choose one object and place it at the edge of a piece of paper.

2. Place your light source on the opposite side of the object to the piece of paper. Do not turn the light on yet!

3. Predict what shape of shadow the object will make when you switch the light on.

   Record your prediction as a drawing on the paper.

4. Switch on the light.

   a Observe the shape that the shadow makes on the paper.

   b How close was your prediction?

   c In a different coloured pencil, draw around the shadow so that you can see both your prediction and the real shadow on the same piece of paper.

5. Repeat this process for the other two objects: predicting, testing and recording.

6. Share your shadow sketches with the class.
   Explain why the objects' shadows are shaped as they are. Use your understanding that light travels in straight lines to help you explain.

Science in Action

Lasers produce a very high-energy, narrow beam of coloured light. Lasers were invented in 1960 and, since then, lasers have been used in many different ways.

Eye doctors use lasers to operate on a person's eye without having to cut the eye open. Lasers let the eye doctor operate very accurately and with less bleeding.

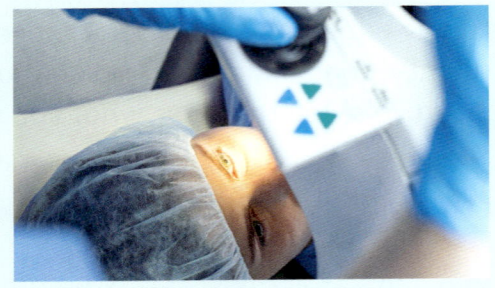

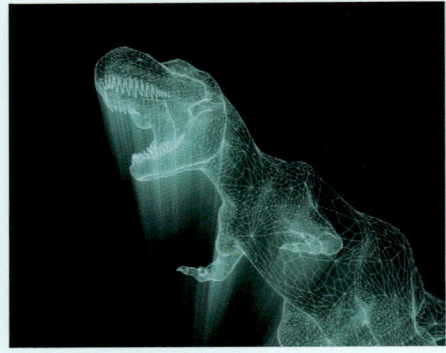

Computer printers also use lasers to print out documents. Holograms use lasers to create images that look as if they are 3-D, but are really only flat.

Traffic police can use laser beams to measure the speed of vehicles. Because the speed that the light in the laser travels at is so fast, the laser beam from the speed detector travels out and back in 'a split second'. This lets police measure the speed of approaching cars and see which ones are driving too fast.

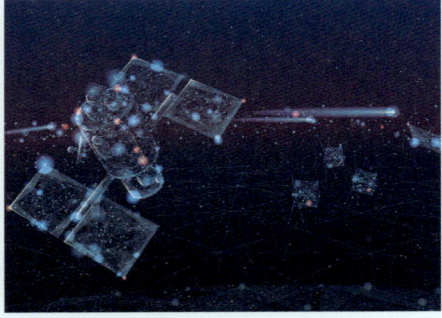

Space communication uses lasers to send data between satellites, spacecraft and the space control centres on Earth.

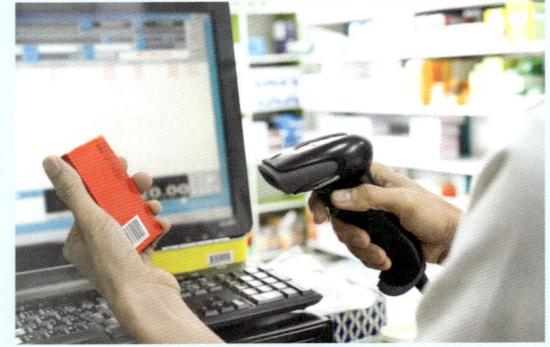

Disc drives in computers and music centres and barcode scanners in shops all use laser light. As time goes by, more and more uses for lasers are being developed.

End of unit check

Choose the correct ending to each of these statements:

1 We can see objects when:

    a light travels from them to our eyes

    b we look at them

    c they are shining

    d they are near us

2 Shadows are created by:

    a transparent objects

    b opaque objects

    c translucent objects

    d all objects

3 Light can be _____ from objects.

    a refracted

    b absorbed

    c deflected

    d reflected

4 You can see over a wall by using a:

    a microscope

    b periscope

    c telescope

    d stethoscope

5 We can see objects behind us by using:

    a glasses

    b mirrors

    c torches

    d goggles

UNIT 5 Electricity

Voltage effects

Activity 1

You will need: a circuit with one bulb and a switch; three cells of the same voltage, paper (or Workbook) and a pen or pencil.

🔍 **1** Look at these **circuits**.

💬 With your group, discuss the differences between them.

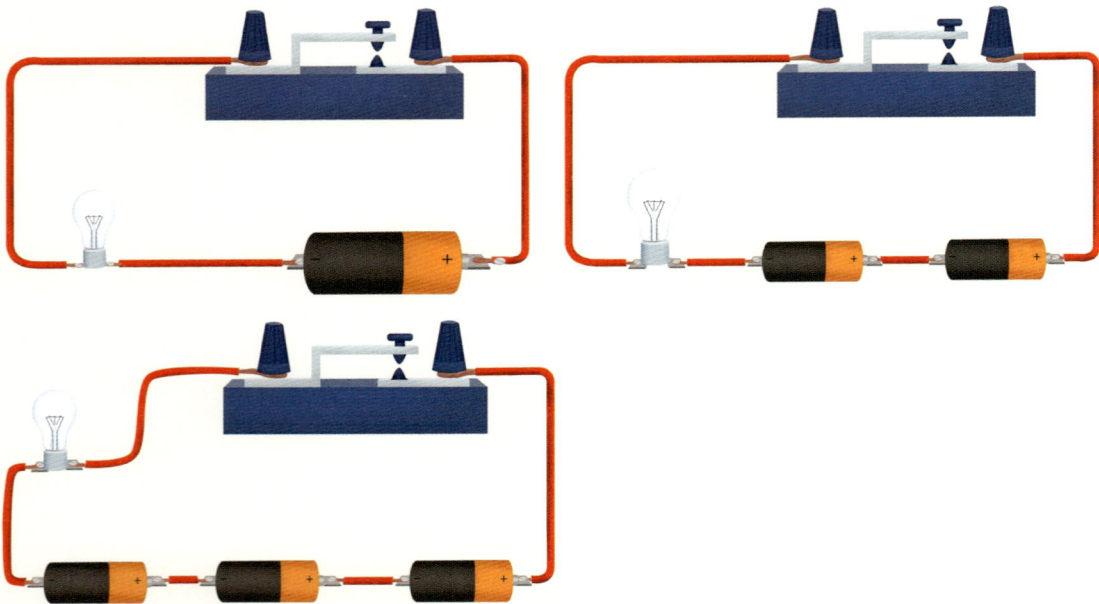

✏️ **2** Predict what you will see if you increase the number of cells (batteries) from one, to two and then to three, and close the switch.

Write down your prediction.

3 Test your prediction.

a Set up a circuit with only one cell. Record the **voltage** of the cell and what you see when you turn on the switch.

➡️ *Workbook page 47*

b Now add another cell to the circuit and turn on the switch. Record the total voltage of the cells and what you see this time.

c Finally, add the third cell to the circuit and turn on the switch. Record the total voltage of the cells and what you see.

4 Compare what you see with three cells to what you saw with one and two cells.

> Make sure that closing the switch is the last thing you do when building a circuit. Once you close the switch, electricity can flow around the circuit. With only batteries like this, the current is so small that you will not get a shock. But it is best to learn this important rule now so that you will stay safe if you work with higher currents in the future!

5 Look at your results and compare them with your prediction. Was your prediction correct? Did your test results prove it right or wrong?

6 Share your results with the class. Try to explain your results and come to a conclusion.

A cell is a store of **energy**. When you connect a cell to a circuit, the energy from the cell pushes the electrical current around the circuit. The size of this 'push' is measured in **volts**. The **current** (which is the flow of electricity) travels out from one end of the cell and back in through the other end.

Adding a second cell doubles the voltage and the amount of energy in the circuit, and so more current flows into the **filament** of the bulb. More current makes the filament hotter. As the filament gets hotter, it produces more light.

A third cell adds even more voltage, more energy and more current to the circuit. The filament gets even hotter and the bulb shines even brighter.

> If you add too many cells to the circuit, the filament becomes so hot it melts, breaks the circuit and the light goes out.

Activity 2

You will need: a switch, a cell, three identical bulbs and five wires, paper (or Workbook) and a pen or pencil.

🔍 **1** Look at these circuits.

💬 With your group, discuss the differences between them.

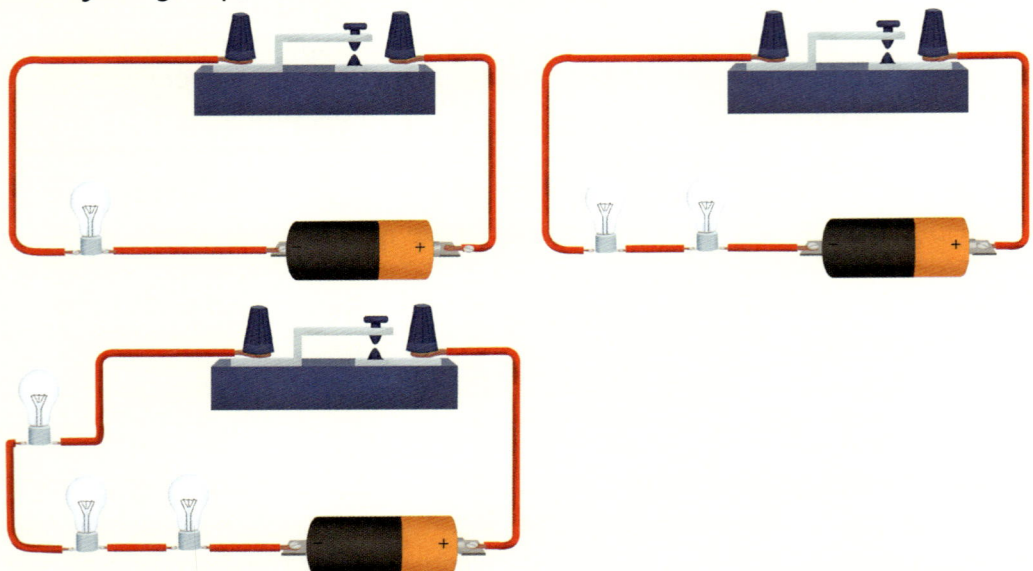

✏️ **2** Predict what you will see if you increase the number of bulbs from one, to two and then to three, and close the switch.

Write down your prediction.

3 Test your prediction.

    a Set up a circuit with only one bulb. Record what you see when you turn on the switch.

    b Now add another bulb to the circuit and turn on the switch. Record what you see this time.

    c Finally, add the third bulb to the circuit and turn on the switch. Record what you see.

4 Compare what you saw with three bulbs to what you saw with one and two bulbs.

5 Look at your results and compare them with your prediction. Was your prediction correct? Did your test results prove it right or wrong?

6 Share your results with the class. Try to explain your results and come to a conclusion.

When you connect bulbs in a line (as they are in the diagrams in Activity 2), we say they are connected in series. Each bulb resists the current – it 'pushes against' the flow of current. So, as you add more and more bulbs to the circuit, less and less current flows because each bulb 'stops' a little bit of the current.

Because there is less current flowing around the circuit, less energy is supplied to each of the bulbs. This means that their filaments do not get so hot, so they give off less light. Adding more and more bulbs in series reduces the brightness of the bulbs.

Circuit components

In Year 4 you explored the differences between materials that allow an electrical current to flow through them – the **conductors** – and those which blocked the flow of current – the **insulators**.

You learnt that all metals conduct electricity, but some conduct better than others.

In this section you will use different types of material to build a circuit and explore what effect this has on the current in the circuit.

Activity 3

You will need: a range of components for you to choose from, including cells, bulbs, wires of different lengths and thicknesses, switches, buzzers, motors; paper (or your Workbook) and a pen or pencil.

In Activities 1 and 2 you investigated how the number of cells or bulbs in a circuit changed the brightness of the bulbs.

Here, you will look at the effect of changing other **components** in a circuit.

1 Choose to investigate either **a**, **b** or **c**.

    a What happens when you change the type of components in the circuit (for example changing from bulbs to buzzers or motors)?

    b What happens when you change the size of components in the circuit, (for example changing from short wires to long wires or changing from thin wires to thick wires)?

    c What happens when you try a combination of both changes?

2 Discuss your choice with your group.

    a Plan a fair test to find out what effects these changes will have on a circuit.

    b Decide what you will observe to find out what effect each change has on the circuit.

    c Decide how you will record your results.

> Remember the value of repeated observations.

3 When the planning is complete, show it to your teacher.

When your teacher is happy with your plan, collect the components you need for your investigation.

4 Before you make each change to the circuit, write down your prediction of what effect the change will have.

5 Carry out the testing in a fair way.

  a Only make *one* change each time. Keep everything else the same, so you know what effect each change has on the circuit.

  b Record the effect of each change *before* you make the next change to the circuit.

6 When you have made all the changes you planned to, look at your results and do these four things:

  a Compare your results with your predictions – which of your predictions was correct? How were your predictions supported by what you observed?

  b Were there any patterns in your results? Were there any results that seem 'odd' and do not seem to fit any pattern?

  c Use your results to make conclusions about how changing the components affects the circuit.

  d Use your scientific knowledge about electricity to explain why each different component affected the circuit in the way it did.

7 Share your results, conclusions and explanations with the class.

Any material that electricity flows through resists the flow of current (the flow of electricity) to some extent. We call this **electrical resistance**.

• Some materials resist the current so much that only a tiny bit can pass through. We say these materials are electrical insulators. They have a very high resistance.

• Other materials, such as metals, hardly resist the current at all. Almost all the electricity passes through them. These materials are electrical conductors. They have a very low resistance.

Any component that you put in your circuit will resist the current to some extent. So, the more components there are, the greater the total resistance and the smaller the current flowing through the circuit.

Some types of component resist the current more than others – a motor has a higher resistance than a lamp, for instance.

The thickness of wires affects the amount of resistance. The length of wires also has the same effect. Using your class investigation of changing circuits, try to answer the following questions.

• Which is more resistant: a thin wire or a thick wire? Why?

• Which is more resistant: a long wire or a short wire? Why?

Dimmer switches use resistance to control how much current flows through the circuit.

When you 'dim' the light, what you are really doing is increasing the resistance in the circuit. This decreases the amount of current that flows through the circuit, and the light gets dimmer.

Drawing circuit diagrams

When we make different circuits, it's important to draw what each circuit looks like so that we can make it again. However, it can take a lot of time to draw all the components in a circuit. Also, different people might draw the same component differently.

Instead of drawing each component as it looks like in real life, we draw a **circuit diagram**. Circuit diagrams use a different **symbol** to represent each component.

Activity 4

You will need: paper (or Workbook) and a pen or pencil.

1 Look at the drawing and the diagram of the same **series circuit**.

   a Compare them.

   b Match the components in the drawing to the symbols in the circuit diagram.

> A series circuit is one where all the components are joined in a ring, one next to the other. There is only one pathway for the electricity. This is the only type of circuit you've learnt about yet, but you will learn about other types in the future.

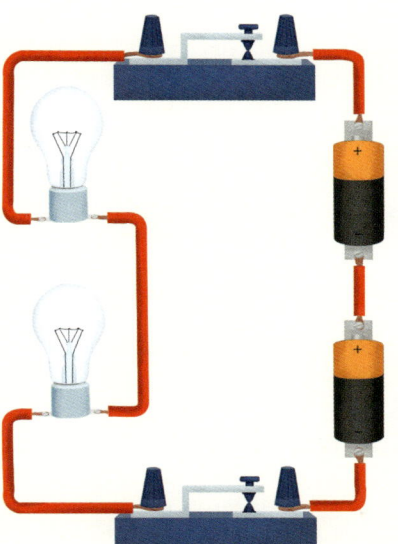
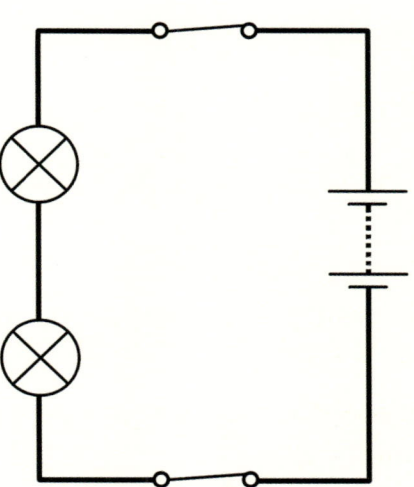

2 Use the same symbols to draw a circuit diagram of a series circuit with three lamps, three switches and two cells.

3 Share your diagram with the class.

Activity 5

You will need: various electrical components from the collection in class; paper (or Workbook) and a pen or pencil.

1 Look at this circuit diagram and work out what each of the components are.

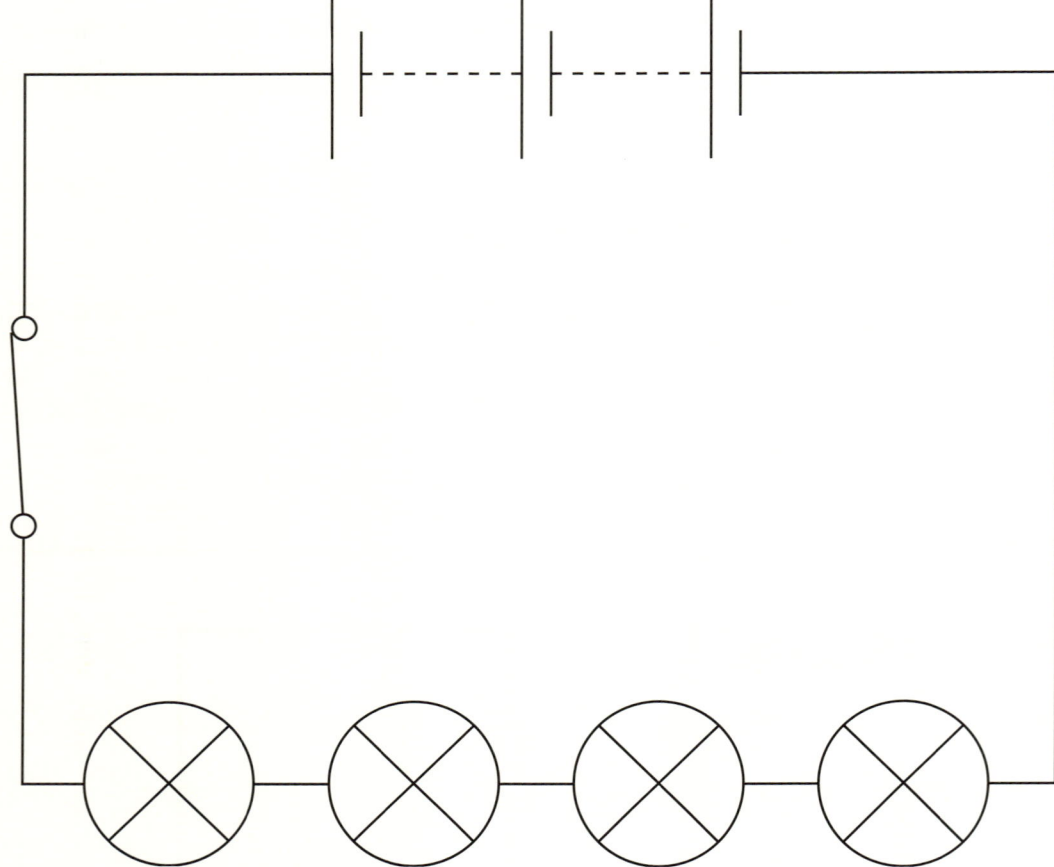

2 Use the diagram to build this circuit using the components you have in class.

3 When the circuit is complete, make a life-like drawing of it (*don't* draw a circuit diagram).

   a Show the circuit and your drawing to the class.

   b Look at the circuits and drawings that other people have made.

Science in Action

The components in a circuit can be damaged if the current becomes too high. The wires can overheat, melt the plastic covering and start a fire.

People can also get bad electric shocks, burns, or even be killed if they touch circuits where the current is too high.

Electric circuits have **fuses** in them to protect us from fires. They protect the circuit and the appliances connected to it. Plugs contain a 'cartridge'-type fuse. Inside a fuse is a metal wire that melts easily if the current becomes too high.

The fuse acts as the 'weak link' in the circuit, so when the fuse is overheated and 'blows', it breaks the circuit and the current stops flowing.

In most modern plugs the fuse has a **rating** of 3 Amps or 13 Amps. It is important to match the fuse to the appliance. Appliances that need more current, such as cookers, heaters and kettles should have a 13A fuse. Items such as TVs, lamps and fridges should be fitted with 3A fuses.

End of unit check

Match the words in Box A with words from Box B. Write down the word pairs.

BOX A
diagram   circuit   switch
current   cells

BOX B
flow   series   symbols
energy   control

_____   _____

_____   _____

_____   _____

UNIT 6 Rocks and soils

Rock types

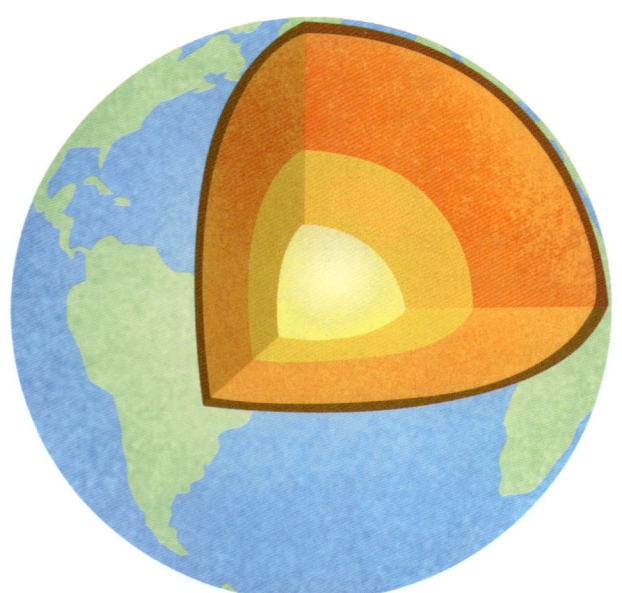

The structure of the Earth

Layers of rock in the Earth's crust

The Earth's **crust** is made of rocks.

In Year 3 you learnt that there are many different types of rock, and that each type has different **properties**.

• Some are hard and others are soft or crumbly.
• Some contain crystals and others do not.
• Some are brighter colours and others are grey or dull.

Rocks are **mixtures** of **mineral** grains that are pushed hard together. A rock's physical properties come from the properties of the minerals that make it.

Scientists sort rocks into the following three types, based on how they were formed.

➡ *Workbook page 60*

Type 1: igneous rocks

Igneous rocks were formed from **lava** that rose from inside the Earth's **core**. It came through the layers of the Earth's surface and some flowed out from volcanoes. When it cooled it formed rocks like **basalt**.

Some lava did not reach the Earth's surface but slowly cooled underground. This made rocks like **granite**. Granite has large crystals of different minerals that are easy to see.

Rocks that cooled down quickly on the surface of the Earth (particularly under the sea), have very small mineral crystals that may not be visible at all without a magnifying glass.

Igneous rocks never contain fossils. Can you explain why not?

Granite

Basalt

Gabbro

Obsidian

Examples of igneous rocks

Activity 1

You will need: paper (or Workbook) and a pen or pencil.

1. Look at the photos of the four igneous rocks. Identify which of these rocks formed:

    a   under the surface

    b   on the surface

2. Tell your group what you think.

    a   Explain your opinions to your group.

    b   Listen to others' opinions and ask them questions.

3. Share the group's ideas with the class.

Type 2: sedimentary rocks

Sedimentary rocks were formed from small particles (called sediments) that settled in layers, either at the bottom of seas, lakes and rivers, or after being blown by winds or carried by ice. This process of sediments settling in layers is called **sedimentation**.

> Sedimentary rocks are generally the type of rocks that contain fossils. Can you explain why?

The particles are a mixture of minerals, often in all shapes and sizes, but sometimes very uniform in composition and size. As the particles built up, the layers below were pressed by those on top of them. This hardened them and the loose particles became cemented together to form layers of rock.

Limestone

Sandstone

Chalk

Siltstone

Examples of sedimentary rocks

The sediments that form sedimentary rocks sometimes come from tiny pieces of igneous rocks in the Earth's crust. At other times, the sediments come from the skeletons of millions of creatures that lived in water and sank to the bottom of seas and lakes when they died. The tiny pieces of their decayed skeletons became squeezed into layers and cemented together by minerals such as quartz and calcium carbonate.

Activity 2

You will need: paper (or Workbook) and a pen or pencil.

1 Look at the photos of the sedimentary rocks. Using the information above, identify the rocks that formed at the bottom of the sea, made up of millions of the skeletons of dead sea creatures.

2 Tell your group what you think and give reasons.

3 Listen to others' ideas and ask questions.

4 Share the group's ideas with the class.

Type 3: metamorphic rocks

The **metamorphic** rocks are the third type of rock. These were formed from rocks that were originally either igneous or sedimentary. The original rocks have been changed – either by heat, or by pressure, or both. They have gone through a **metamorphosis** – which gives them their name.

The change usually takes place deep under the surface of the Earth and so these rocks are only seen when the layers above them are removed in some way, such as by **erosion**.

> Metamorphic rocks have often been heated and put under very great pressure. This usually breaks any fossils up. So metamorphic rocks rarely contain any fossils.

Slate is made from the sedimentary rock shale.

Gneiss is made from the igneous rock granite.

Marble is made from the sedimentary rock limestone.

Schist is made from the sedimentary rock mudstone.

Examples of metamorphic rocks

The hardness of rocks varies a great deal according to the minerals in the rock. Scientists have arranged minerals in order from the softest, talc (Number 1) to the hardest, diamond (Number 10).

Activity 3

You will need: samples of igneous, sedimentary and metamorphic rocks, a hand lens, a nail, paper (or Workbook) and a pen or pencil.

1. Examine each of the three rock types to observe their **composition** and structure.

   a Record your observations in a table, listing and describing the characteristics of each rock, for example colour, texture, hardness and whether it is dull or shiny.

   b Compare the hardness of the rocks by trying to scratch each one with the nail. Draw and colour a picture of each rock.

2. Go outside and collect at least three different types of rock. Look for stones and pebbles on the ground and in the soil. If there is a place where rocks are exposed – such as a river bank or by the sea – collect rocks from there too.

3. Back in class, compare the rocks you collected with those you have already examined.

   a Try to classify each one as either igneous, sedimentary or metamorphic.

   b Write your observations about the rocks you collected in the table of characteristics you made in question I.

4. Display the rocks you found. Sort them into sets of igneous, sedimentary and metamorphic rocks.

5. Look at other people's sets of rocks and compare them with yours.

6. Share your ideas in the class discussion.

The rock cycle

Changes in the Earth's crust

The Earth's crust changes all the time. Most of these changes are slow and we are not aware of them. Here are some of the ways that these changes happen.

Rocks are worn away by **weathering**. Rain, wind and frost all slowly remove particles of rock from exposed surfaces. The rock wears away, little by little, as it loses material from the outside. These changes are very slow, but over long periods they change the rocks on the surface of the Earth.

Chemical weathering also wears rocks down if there is **pollution** of the air or the water. Acid rain is an example where rainwater is changed into a weak form of acid by chemicals in polluted air. This acid attacks the rocks when it lands on them, wearing them away more quickly.

Rocks are **eroded** by rivers or glaciers. Rivers often carry **sand** or **silt** in them as they move over the surface of the rocks. These solids scrape away at the rocks as they move over them.

Gorges, canyons and other deep valleys can be created by rivers eroding the rocks over a long time. They also produce waterfalls in some places. Glaciers on the high mountains and at the South Pole scrape rocks as they slowly slide over them. They can carry the material that they break off over long distances and it is deposited eventually when the ice melts.

Erosion at the coastline takes place all the time. It is faster in some places than others, because some rocks are harder than others and the sea is more powerful in some places than in others. The sea can carry the eroded material along the coast and drop it in a different place, or it may be taken out to sea, where it falls to the seabed.

Sometimes the changes are sudden and violent! Earthquakes and volcanoes can tear the rocks in the Earth's crust apart and bring new ones to the surface.

The Earth's crust is not a single covering layer, like the skin of an orange. Instead, it is divided into sections of various sizes called **tectonic plates** that shift and push against one another.

Deep in the oceans there are cracks where two tectonic plates meet. Molten lava from the core leaks out of these cracks all the time and forms new rock on the sea bed.

Sudden shifts along the edges of the plates, where they push against each other, can cause earthquakes and **tsunamis**.

New rocks being formed on the sea bed

The rock cycle

The rocks of the Earth are constantly being recycled. They change from one type to another, they move lower and higher in the crust, they are exposed and buried, eroded and built up. All these movements and changes are called the **rock cycle**.

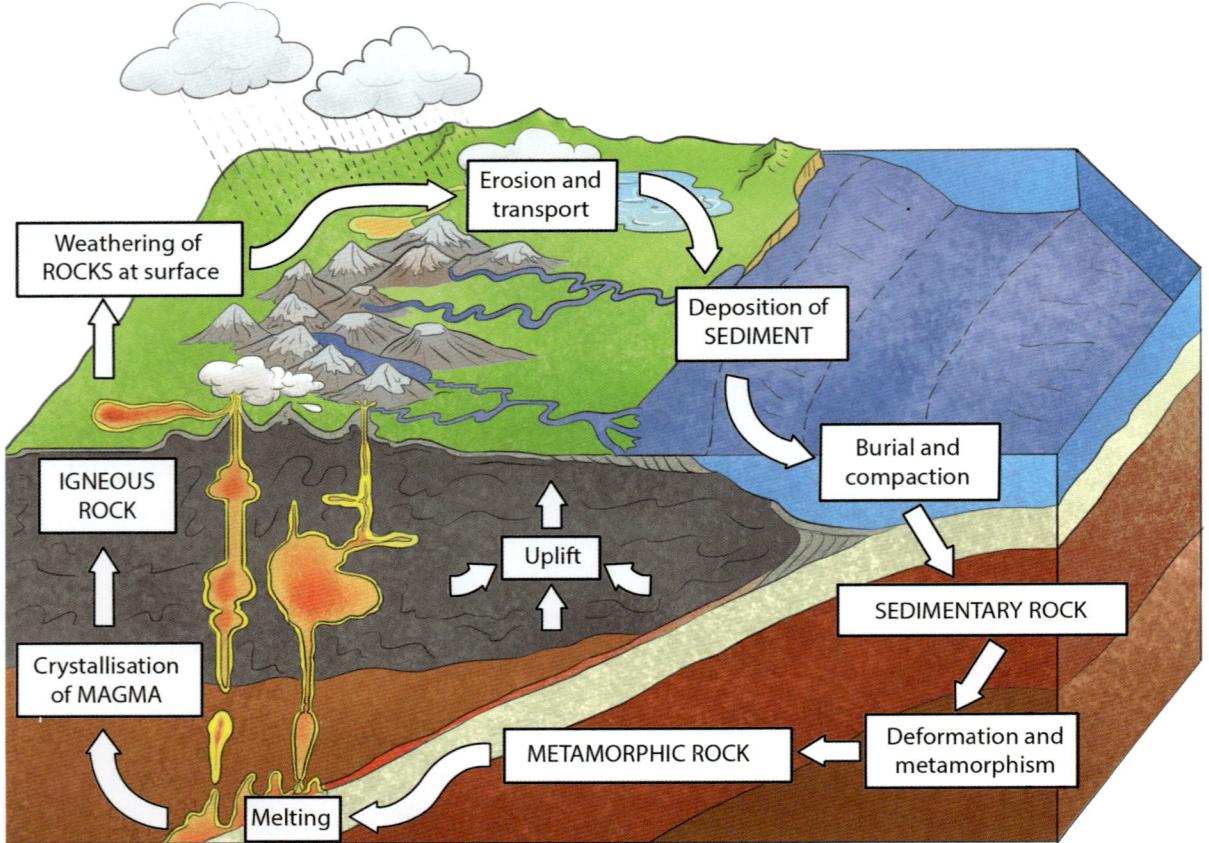

The rock cycle

Here are some key parts of the rock cycle. Try to identify each one in the diagram.

- Igneous rocks can be eroded and the particles end up as sediment, starting the formation of new layers of sedimentary rocks.
- Sedimentary rocks can get heated or squeezed and can be transformed into metamorphic rocks.
- Metamorphic rocks, which once solidified and formed part of the crust, can be melted again when they get buried and pushed down towards the core.
- The melted metamorphic rocks can form igneous rocks when they cool at the Earth's surface.

Soil types and plant growth

In Year 3 you explored what soil is made of. You found that soils, like rocks, are mixtures.

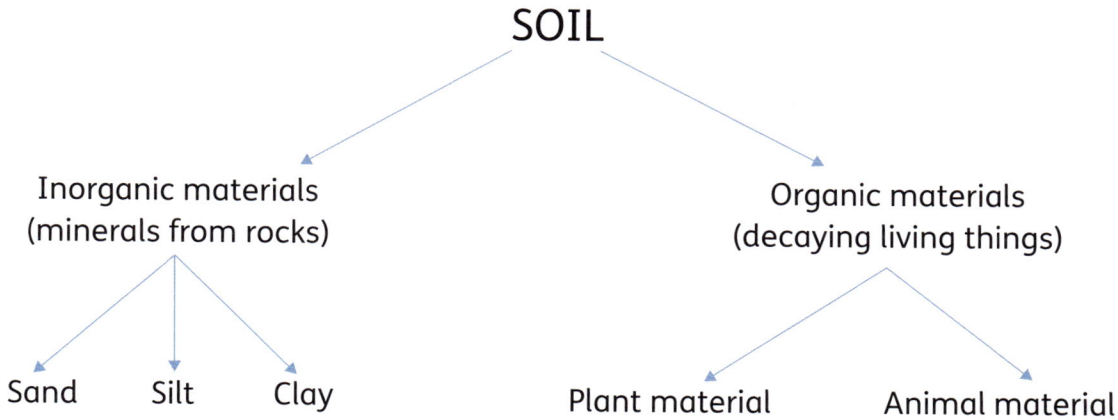

What soil is made of

Each type of soil contains **organic material** (that once came from plants or animals) and **inorganic material** (minerals that were once part of rocks). This is why soils are so varied.

The organic part of soil (from decaying plants and animals) is called **humus**. Humus contains nutrients that plants need for healthy growth.

In Year 3 you learnt to separate the different inorganic minerals in soil by adding water. Sand and gravel are the largest mineral particles and **clay** is the smallest. In Year 3 you mixed soil with water and let it settle. The larger particles settled first and the extremely small clay particles took hours to finally sink to the bottom.

Activity 4

You will need: two tins of the same size, two bowls of the same size, two jars or beakers of the same size, water, dry sand and dry clay, a ruler, paper (or Workbook) and a pen or pencil.

In this activity you will use a fair test to see whether water drains more quickly through sand or through clay.

1 Your teacher will help you by making some holes in the bottom of both tins. The number of holes, and the position of the holes, should be the same for both tins.

> Why do the holes need to be the same in both tins?

2 Fill the tins.

  a Label one tin 'sand' and the other tin 'clay'.

  b Use the ruler to make a mark on the inside of each tin 3 cm below the top of the tin.

  c Fill one tin up to the mark with sand and the other tin up to the mark with clay.

> Why does the amount of sand need to be the same as the amount of clay?

3 Write down your prediction of what will happen when you pour water into the two tins.

  a Will the water drain through?

  b Will the water drain through at the same speed in both tins? Which will be quickest?

4 Pour the same amount of water into the two jars or beakers. Why does this need to be the same?

5 Pour water through the tins.

   a Two members of the group should hold the tins above the two bowls.

   b Another two group members should pour the same amount of water into the tins at the same time.

Why does it need to be at the same time?

6 Observe what happens.

   a Discuss with your group what you can see as it happens.

   b Continue to observe until the water has stopped coming through the holes of both tins.

7 Pour the water from each bowl back into the jar or beaker you poured it from.

Compare how much water has drained through each tin.

8 Write down what you have observed.

   a Compare the result with your prediction.

   b In your group, try to explain *why* your results show what they do.

What was the difference between the sand and the clay?

9 Share your results with the class.

Clay and sandy soils

Water can move through soils because it is a liquid and the air spaces (**pores**) between the mineral particles allow it to move into and through the soil. Where the pores are big, as in sandy soils, the water can move through easily. In clay soils, the pores are extremely small and water gets trapped.

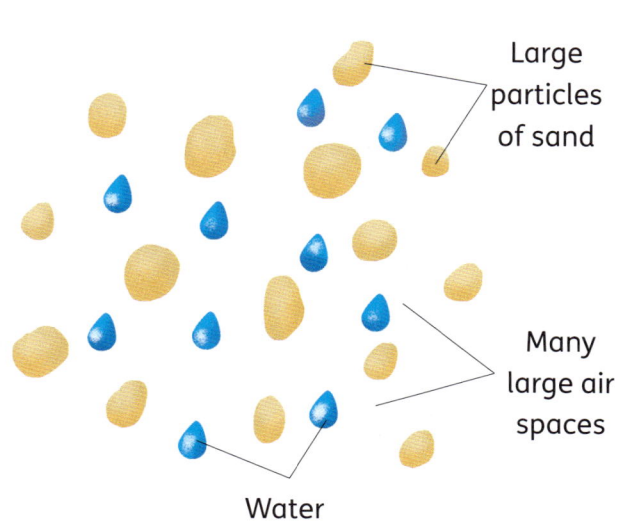

Large particles of sand

Many large air spaces

Water

Sandy soils have a lot of air in them, but they do not hold on to water. It drains through easily and this means that plants in sandy soil often cannot find enough water to stay alive. Dry, sandy soil is also easily blown or washed away.

Clay soils are very good at holding water. However, soil with too much clay is also bad for plants. When it rains, clay soil becomes wet and sticky. The water pushes the air out of the soil and without air the roots die. When clay soils are dried out, they bake hard like bricks.

Soils need to contain a mixture of sand and clay so that the plants get enough water and oxygen.

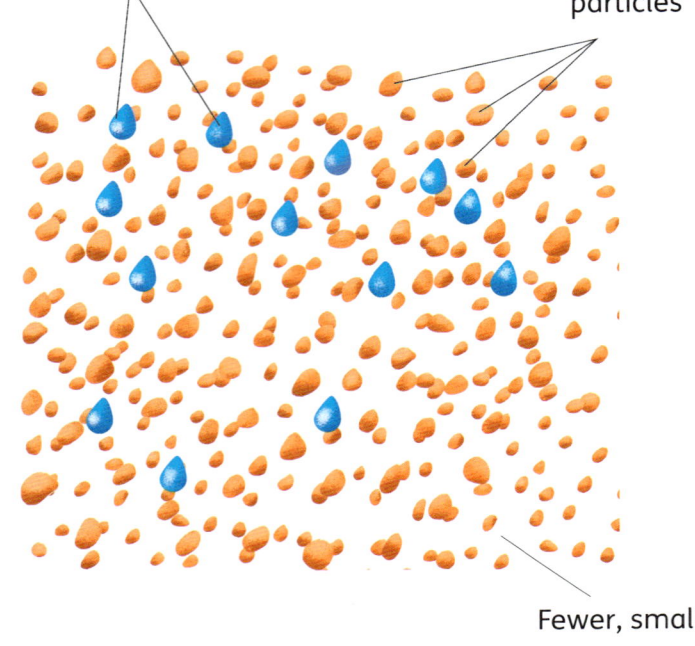

Water

Small clay particles

Fewer, small air spaces

Loam soil

The best soil for plants is called loam.

This table shows the mixture of materials found in loam.

Organic material	The inorganic material in loam soil is about 40% sand, 40% silt and 20% clay		
Humus	Sand	Silt	Clay
Humus contains nutrients that plants need for healthy growth.	The sand and silt in loam soil means that water drains through freely so that plants have enough air.		The clay in loam soil holds onto some water for the plants to absorb.

Science in Action

Farmers and gardeners are always interested in how much clay, sand and humus are in their soils. This can have a very good or bad effect on how well their flowers or crops grow.

Different crops can grow better in slightly different types of soil. Most crops need a mixture of humus, sand and clay, but the quantities of each might vary.

Carrots grow best in deep, sandy loam soil.

Broccoli grows best in clay loam soil.

Flowering plants often need soil that is rich in humus.

Finding the right soil is very important to plants and crops if they are to grow well.

When crops fail to grow, people go hungry, suffer famine and can die. World food production is the most important 'industry' of all. We can live without many things we like, but we cannot live without food.

We all need soil to keep us alive and healthy.

End of unit check

Use the words from the box to complete the sentences.

seabed	sand	sedimentary	easily	rocks	weathered	rain
core	chalk	sedimentary	cycle	clay	igneous	size	soils
liquid	cools	Earth's	changing	metamorphic	heating

1. _____ can be sorted into three types: metamorphic, igneous and _____.

2. _____ rock comes from the Earth's _____ to the crust and _____ down, forming _____ rocks.

3. _____ rocks are made by _____ or squeezing other rocks.

4. Some _____ rocks, such as _____ and limestone, were formed on the _____ and in lakes.

5. The _____ crust is constantly _____ and rocks are being made and changed in different ways. This whole process is called the rock _____.

6. Rocks can be _____ by heat, _____ and wind.

7. The rock particles in soil are called _____ , _____ or silt depending on their _____.

8. Sand helps water to _____ drain through the soil so that plants get enough air. Clay _____ holds the water so that plants get enough to absorb.

Glossary

adaptation – how living things are specialised to suit their environment.

addictive substance – a substance that makes the user dependent upon it (for example alcohol, nicotine).

alcohol – a colourless, poisonous, liquid drug found in alcoholic drinks.

aluminium – a silvery-white metal very common in the Earth's crust.

anus – the hole at the lower end of the gut, through which the faeces pass out of the body.

artery/arteries – blood vessels (tubes) that take blood from the heart to all parts of the body.

bacteria – micro-organisms that can have beneficial effects (for example in cheese making or can cause diseases such as cholera and tuberculosis).

balanced diet – a diet that contains some of all the food types the body needs.

basalt – a type of volcanic rock that cooled rapidly on the surface of the Earth.

bile – a liquid made in the liver that is added to food in the intestine; it helps us to digest fats.

bladder – the elastic bag in which urine is collected before being released from the body.

blood vessels – tubes along which blood travels to and from all parts of the body, such as arteries, veins and capillaries.

botanist – a scientist who studies plants.

brain – the organ that responds to, remembers and is aware of the environment outside and inside the body; the organ of control, learning and personal expression.

capillary/capillaries – the smallest tubes that carry blood between the cells.

carbohydrates – nutrients in the diet needed for energy (for example sugars, starch).

carbon dioxide – a gas in the air, used by plants to make food; we make carbon dioxide in our bodies and breathe it out.

carnivore – a consumer that feeds on other animals (for example cat, tiger, snake).

cell – (biological) a small unit of which tissues and whole bodies are composed; (electrical) a source or store of electrical energy, often called a 'battery'.

characteristics – features that are always found; typical things.

chemist (pharmacy) – a shop where medicines can be bought (prescription and over-the-counter drugs).

circuit (electrical) – a complete circular route around which electricity can flow.

circuit diagram – a drawing made up of symbols for the components of a circuit.

circulatory system – the heart, vessels and blood, which distribute food, hormones and oxygen and remove waste products from the body's cells.

classification/classify – sorting living things into groups based on common features (for example all creatures with a backbone are vertebrates).

clay – the smallest mineral particles of many soils.

compare – to look for differences and similarities in two or more things or events.

competitive – able to survive by being better at obtaining things needed for life (for example food, light, water).

components – parts.

composition – what something is made of, the ingredients, the parts.

conductor (electrical) – a material through which electrical energy (electricity) is transferred or flows.

consumer – an animal that eats plants or other animals, or both, as food.

core – the centre of the Earth where rocks are liquid and very hot.

cornea – the transparent 'window' that covers the front of the eye.

crust – the solid, outermost layer of the Earth.

current – a flow of electrically charged particles.

diabetes – a disease that involves problems with sugar levels in the blood.

dialysis – a treatment for a person with kidney or, sometimes, liver failure, that passes their blood through a machine to remove poisonous wastes (for example urea).

diet – all the things we (and animals) eat and drink.

digestion – the process of breaking food into small, soluble substances (nutrients) that can be absorbed by the body.

digestive system – the stomach and other parts inside the body that turn food into useful substances (nutrients) needed by the body.

dimmer switch – a switch that can vary the brightness of a lamp.

donor heart – a heart from someone who has died that is put into another person.

dose – an amount of a drug taken at one time.

drug – any substance, other than food, that causes a change in the body when swallowed, breathed in, injected or applied to the body (for example tobacco, alcohol, antibiotics and painkillers).

drug abuse – using drugs for the wrong purpose, or in doses that are excessive.

electrical resistance – the resistance of a material to the flow of an electric current passing through it.

energy – the ability to do work. It is needed to make things happen.

environment – physical surroundings, including the weather.

eroded – worn away by physical or chemical forces.

erosion – wearing away; the process that helps to turn rocks into soils and breaks down coastal cliffs.

evolution – the theory that a process of gradual change occurs in each generation of living things.

excrete – get rid of waste products from the body (for example urine, carbon dioxide).

excretion – the process of getting rid of waste products from the body (for example sweating, urination, breathing out).

excretory system – organs (for example kidneys, skin) that remove waste products from the body.

exercise – activity of the body for health and fitness.

extinct – no longer existing, wiped out.

faeces – the solid waste left over after digestion of food, which passes out of the body through the anus.

fair test – a test of an idea in which everything is kept the same except the one thing you are testing.

fats – nutrients used by the body to provide energy – contained in plant oils and animal fats. Fats contain carbon, oxygen and hydrogen.

Felidae – a family of mammals, the cats.

filament – the thin wire in a lamp that gets hot and makes light.

food chain – the sequence of plants and animals that eat or are eaten for food. Each food chain starts with plants.

fossil – remains or evidence of living things from long ago found in rocks.

fossilised – turned into a fossil.

fruits – the part of plants in which the seeds develop.

function – a job, a purpose, a use; work done by something.

fungi – a large group of living things that are not plants or animals. They are useful in decomposing dead organisms. Some are harmful to plants and animals.

fuse rating – the size of the current that can flow through the fuse without it 'blowing', (that is, without the wire in the fuse melting and cutting off the current).

fuses – safety devices found in plugs and electrical appliances that can break the circuit.

granite – a type of volcanic rock that cooled slowly below the Earth's surface and is a mixture of crystals.

habitat – the environment in which an animal or plant normally lives.

heart – the organ that pumps blood round the body, through the blood vessels; part of the circulatory system.

herbivore – a consumer that feeds on plants (for example caterpillar, cow).

hormones – chemicals made by various glands in the body; they control body processes (for example sexual development).

humus – organic matter from plants and animals found in soils.

ice age – period when ice covered large areas of the Earth's surface; the most recent ice age is thought to have ended about 10,000 years ago.

igneous rocks – rocks that were formed from the liquid material in the core of the Earth.

illegal (drugs) – drugs that the law does not allow people to take.

infection – an illness or disease, caused by some organism that has invaded the body.

inheritance – the process of passing on features from parents to offspring.

inorganic material – mineral material, made from rocks.

insulator (electrical) – a material that blocks the flow of an electrical current (for example plastics, wood).

intestine – the part of the digestive system where most of the processes of digestion and absorption takes place.

invertebrate – an animal that does not have a backbone (for example worm, snail, insect).

investigate/investigation – to search for evidence to answer a question.

kidneys – a pair of organs in the abdomen that remove waste products from the blood and pass them on in urine to the bladder; part of the excretory system.

laser – a device that creates a beam of intense light.

lava – molten rock in the Earth's core.

life cycle – all the stages of an organism's life arranged in order.

liver – a large organ above the stomach that has many functions (for example processing, control and storage of nutrients; handling some waste products) part of the digestive system.

loam – soil with balanced amounts of inorganic and organic matter; garden soil.

lungs – a pair of organs in the chest, used to pull air (including oxygen) into the body and to push out the waste gas carbon dioxide, made by the body; part of the respiratory system.

metamorphic rocks – rocks that have been changed by pressure, heat or chemical action, (for example slate, marble).

metamorphosis (rocks) – the process of changing one rock type into another by heat or pressure.

microscopic – so small it cannot be seen without a magnifying lens.

mineral – a substance in rocks and soils that both plants and animals must have for their healthy growth.

mixture – two or more materials mixed together; there is no reaction between them.

mollusc – a soft-bodied animal. Most molluscs have shells (for example garden snails).

muscle – body tissue that can contract and relax. Bundles of this tissue are attached to bones and produce movement. Muscles are also vital parts of other organs (for example the heart, stomach, anus, eyeball).

nervous system – the organs that provide the body with sensitivity and control – brain, spinal cord, nerves and sense organs.

nicotine – an addictive drug found naturally in tobacco leaves.

nutrient – a substance in food needed by the body for various processes (for example growth, control of body temperature, protection).

observe/observation – to notice when paying careful attention (see, smell, hear, touch, taste).

oils – types of fats produced from plants and some animals (for example palm oil, cod liver oil).

omnivore – a consumer that feeds on plants and animals. (Humans are omnivores, for example.)

opaque – light does not pass through and it is not possible to see any image at all through such a material (for example stone).

organ – a part of an animal or plant that has a particular function (for example brain, flower, liver, skin).

organic material – plant and animal material (humus); dead parts, dung, droppings.

palaeontologist – a scientist who studies living things from long ago, using fossils, frozen bodies and other remains.

parasite – an animal, plant, fungus, virus or bacteria that feeds on another living organism (for example worms in the intestine, fungus on the feet, malaria parasite in the blood).

periscope – a device using mirrors that allows a view of objects which cannot be seen directly.

petals – flower parts surrounding the reproductive organs; they are often large, colourful and attractive to insects.

pharmacy (chemist) – a shop where medicines can be bought (prescription and over-the-counter drugs).

photosynthesis – the process in the green parts of plants that makes food from carbon dioxide, water and light.

plasma – the liquid part of the blood in which all the solid parts are floating.

platelets – small particles in the blood that help it to clot when the body is wounded.

pollution – the spoiling or damaging of the environment by harmful wastes.

pores – spaces, holes.

predator – an animal that kills other animals for food.

predict/prediction – to say what will happen before doing something.

prescription – a drug used as a medicine, which a doctor has chosen as the treatment for an illness.

prey – animals (for example mice) that are killed and eaten by other animals.

producer – a plant, which produces its own food through photosynthesis.

product – what is made; the result of a reaction.

prohibited – against the law; illegal.

properties – features, characteristics, of what something is like (for example size, hardness).

proteins – nutrients needed for growth, repair and many basic body functions.

protozoa/protista – single-celled animals.

pupil –the hole at the front of the eyeball through which light enters the eye.

rate – the pace at which something takes place (for example miles travelled per hour of time).

rating (fuse) – the size of the current that can flow through the fuse without it 'blowing' (that is, without the wire in the fuse melting and cutting off the current).

ray – a beam of light; a narrow line or column of light travelling from a source or reflected from a surface.

record – a written account, photo or drawing of what was done or what happened.

red blood cells – most cells in the blood, which carry oxygen to all the cells of the body.

reflect – to bounce back. Light hitting a surface is more or less returned (reflected) or absorbed.

reproduction – the process in living things that produces new individuals.

respiratory system – the organs that supply the body with oxygen and expel the waste gas, carbon dioxide. The nose, mouth, windpipe and lungs form the respiratory system.

results – observations of all kinds, including measurements, collected during an investigation.

rock cycle – the constant creation and destruction of rocks occurring in the Earth's crust.

sand – a material formed from the erosion of rocks that forms a large part of many soils.

saprophyte – an organism, such as a fungus, which absorbs its food from dead organic matter.

sedimentary rocks – rocks formed from particles of older rocks or organic matter.

sedimentation – the process of building layers of material one on another on sea and lake beds.

segment – a part, section.

sense organs – the body parts that have the senses: eyes, ears, nose, tongue, skin.

sepals – the outer part of a flower, often green in colour and covering the petals when the flower is closed in the bud.

series circuit – a circuit where all the components are joined in a ring, one next to the other; there is only one pathway for the electricity.

silt – soil particles larger than clay and smaller than sand.

skin – the outer covering of the body; it has several functions, including excretion and control of body temperature through sweating.

species – a type of animal or plant that is like all others of the same species and can reproduce with members of that group/species.

stamen – the male part of a flower consisting of a filament (stalk) and anther.

stigma – the female part of a flower that is sticky and traps pollen grains.

stomach – a large elastic organ into which food and drink pass when swallowed. Digestion, storage and mixing of the food take place in it. The stomach is part of the digestive system.

sweat – salty liquid excreted by the skin, which evaporates and cools the body.

sweat gland – a small organ in the skin that makes sweat and passes it out via pores.

symbol (electrical) – a sign used to represent a component of an electrical circuit (for example switch, cell, lamp).

tectonic plates – sections of the Earth's crust that can move.

tobacco – dried leaves of the tobacco plant, containing the drug nicotine; they are made into cigarettes and cigars and processed for use in pipes. Smoking tobacco damages health.

transplant – to take an organ from one person and put it into another person (for example kidney, heart) from one person and put it into another person.

tsunami – a series of enormous waves, created by an underwater earthquake or volcanic eruption.

urea – a waste product produced by the kidneys that is disposed of in the urine.

ureter – the tube that carries urine from the kidney to the bladder.

urinary system – the part of the excretory system that rids the body of urine, consisting of kidneys, ureters, bladder and urethra.

urine – a pale yellow liquid excreted by the kidneys to remove waste products from the body.

variable – a factor that can change or be changed (for example temperature).

variation – the differences between living things in a species.

vegetable – (in general) any part of a plant; (in cooking) plant parts that are not fruits or seeds (for example cassava, spinach).

vein – a blood vessel (tube) that carries blood from all parts of the body back to the heart and the lungs.

vertebrate – an animal with a backbone.

virus – a parasitic micro-organism that can cause disease in plants or animals (for example flu).

vitamins – nutrients that are essential to good health, protective nutrients. Vitamins are only needed in small amounts.

volt – the unit of electromotive force.

voltage – a measure of the force of electricity.

weathering – various processes that cause rocks to break down into smaller and smaller particles.

white blood cells – a solid part of the blood that defends the body by attacking germs.

woolly mammoth – an extinct type of elephant that lived in northern areas.

zoologist – a scientist who studies animals.